AF756335

Biomass Energy Systems and the Environment

Pergamon Titles of Related Interest

Goodman/Love BIOMASS ENERGY PROJECTS
Schlegel/Barnea MICROBIAL ENERGY CONVERSION

Related Journals*

ANNALS OF BIOMEDICAL ENGINEERING
THE ENVIRONMENTAL PROFESSIONAL
ENERGY
RECLAMATION REVIEW

*Free specimen copies available upon request.

Biomass Energy Systems and the Environment

Helen M. Braunstein
Paul Kanciruk
R. Dickinson Roop
Frances E. Sharples
Jesse S. Tatum
Kathleen M. Oakes
with contributions from
Frank C. Kornegay
Thomas E. Pearson
Oak Ridge National Laboratory

PERGAMON PRESS
New York ▪ Oxford ▪ Toronto ▪ Sydney ▪ Paris ▪ Frankfurt

Pergamon Press Offices:

U.S.A.	Pergamon Press Inc., Maxwell House, Fairview Park, Elmsford, New York 10523, U.S.A.
U.K.	Pergamon Press Ltd., Headington Hill Hall, Oxford OX3 0BW, England
CANADA	Pergamon Press Canada Ltd., Suite 104, 150 Consumers Road, Willowdale, Ontario M2J 1P9, Canada
AUSTRALIA	Pergamon Press (Aust.) Pty. Ltd., P.O. Box 544, Potts Point, NSW 2011, Australia
FRANCE	Pergamon Press SARL, 24 rue des Ecoles, 75240 Paris, Cedex 05, France
FEDERAL REPUBLIC OF GERMANY	Pergamon Press GmbH, Hammerweg 6, Postfach 1305, 6242 Kronberg/Taunus, Federal Republic of Germany

Library of Congress Cataloging in Publication Data
Main entry under title:

Biomass energy systems and the environment.

Includes bibliographies and index.
1. Biomass energy--Environmental aspects.
I. Braunstein, H.M.
TD195.B56B56 1981 333.95'3 81-10609
ISBN 0-08-027194-4 AACR2

Printed in the United States of America

Contents

Foreword

This work was originally prepared to examine the environmental implications of biomass systems as an energy source and to encourage the incorporation of environmental values into the earliest stage of planning and decision making. Because biomass energy systems comprise both production and conversion technologies that cover a multitude of options and combinations, they penetrate deeply into existing agricultural, economic, and environmental systems. But because they are also conformable, implementation of this diverse technology is a challenge yet to be fully assessed.

This book carries with it a debt of gratitude to several persons who contributed not only to the successful completion of this project, but without whom this work would not have evolved in its present form. Harry G. Arnold, of the Oak Ridge National Laboratory (ORNL) Energy Division, was such a key figure. As program manager during the original organization and subsequent preparation of the draft, he was responsible for coordinating into a systematic and meaningful study what at first appeared to be an almost unmanageable subject area. In addition, this book would not have been possible without the personal interest, dedication, and continued management support provided by the following people: Ray Brechbill, who managed the project through the San Francisco Operations Office; R. B. Craig, who manages the Environmental Impacts Program in the ORNL Environmental Sciences Division; and T. H. Row and H. E. Zittel, who manage the Environmental Impacts Section in the ORNL Energy Division.

Finally, special thanks are due Myrtleen Sheldon, of the ORNL Technical Publications Department, who directed all phases of the editorial work preparatory to publication. The authors are also indebted to many people, too numerous to list, who reviewed the manuscript and offered comments and suggestions. Significant comments were received from members of the ORNL Environmental Sciences Division, and helpful suggestions were received from scientists in the Tennessee Valley Authority Biomass Program. Last, but not least, thanks are due managers and members of the many organizational units of ORNL who contributed visibly or otherwise to the successful completion of this book.

H. M. Braunstein for all the authors
January 1981

1
Introduction

Solar energy, converted to chemical energy by photosynthesis, is stored in a variety of organic materials generically termed "biomass." As used in this book, biomass includes wood, herbaceous plants, crops, residues from agricultural and forest products, manure, freshwater plants, and marine plants. Biomass, primarily as wood, was the fuel that provided the major portion of mankind's energy needs until the last century. Recently, the national search for alternatives to fossil fuels has rekindled interest from many sectors of the economy in biomass as a source of energy. National policy encourages the development of liquid and gaseous fuels from wood, crops, crop residues and manure to displace a portion of the country's imported oil. Each year, more residential consumers are turning to wood to supply some portion of their home-heating needs. Industrial and agricultural sectors are examining the reliability of various biomass materials as economical sources of power and heat. Biomass could also provide a significant part of the energy for developing nations faced with a future of scarce supplies of petroleum. Biomass energy systems are rapidly becoming a promising technology for the conversion of solar energy. The time is therefore ripe for an examination of the environmental and social implications of a large-scale commitment of biomass resources to the production of energy.

This book provides basic information on biomass energy systems to scientists, managers, businessmen, administrators, and the general public, with emphasis on an examination of environmental and social issues related to producing energy from biomass. Biomass energy systems encompass a wide range of production technologies, feedstock materials, and methods for conversion to fuels. This study is structured to reflect the two distinct phases of biomass energy systems: production and harvest of biomass feedstocks and conversion of these feedstocks to energy. The chapters related to biomass production reflect the five sources of biomass—forests, herbaceous crops, livestock wastes, freshwater plants, and marine plants. Similarly, the chapters devoted to conversion reflect the four categories of technologies for conversion of biomass to energy—anaerobic digestion, fermentation, thermochemical conversion, and direct combustion.

The book is organized into seven chapters, including this Introduction. Chapter 2 describes the existing environment and resource base for the production of biomass, emphasizing current uses of land, water, and energy and present uses of biomass materials. Chapter 2 also examines the environmental effects of current levels of biomass production to provide a baseline for later discussion of incremental effects from production of biomass for energy. Chapter 3 describes the schemes for production and harvest of biomass feedstocks, and Chapter 4 describes the technologies for conversion of the biomass feedstocks to energy. Some of the processes described in these two chapters are well known; many others are in the conceptual or early design stages. The last three chapters of the book are devoted to discussions of potential environmental and social consequences of the implementation of biomass energy systems on a national scale. Chapter 5 discusses environmental impacts of production and harvest of biomass feedstocks, Chapter 6 examines potential effects from conversion technologies, and Chapter 7 describes social and economic issues related to the use of biomass for energy production. Whenever possible, predicted effects are quantified; however, because of the many variables involved in each biomass-to-energy cycle and because most of the technologies for large-scale production and conversion have yet to be demonstrated, quantification, confident predictions, and comparisons among biomass technologies were not always feasible.

Biomass energy technologies are popularly regarded as "soft" technologies—small, simple, dispersed, renewable, and environmentally benign. This view is naive; there are important and complex concerns related to large-scale exploitation of biomass for energy. The more critical concerns include competition with food and fiber crops for scarce land and water resources, disturbance of natural ecosystems, increased soil erosion, degradation of air and water quality, encroachment on wildlife habitat and wilderness areas, and alteration of the social and economic character of rural communities.

The greatest threat to the environment could come from production of biomass feedstocks to be converted into energy. Experience from the agricultural and forest products industries has shown that irresponsible production and harvest techniques can damage not only the quality of the surrounding environment, but also that of the production resource base itself. Sunlight is diffuse, and green plants convert into carbodydrates only about 1 to 2% of incident solar energy. Use of plants to concentrate even moderate amounts of energy will require large commitments of land, water, and fertilizer. Areas of the nation where the combination of climate, topography, soil, and availability of water is favorable for the production of biomass are already largely devoted to food and fiber production. Furthermore, heavy exploitation of this optimal resource base has already caused a decline in its productivity. Two major questions about the future of biomass energy therefore become: (1) Can the production resource base be shared so as to meet the needs of both food, and fiber and energy production? (2) Can large-scale production of biomass for energy avoid intensifying the environmental effects characteristic of existing food and fiber production?

At the conversion end of the biomass energy cycle, potential environmental degradation can ensue from many uncontrolled small-scale uses of biomass to produce energy. Residential woodburning provides the best illustration of this concept. Recent experiences in Aspen, Colorado, Portland, Oregon, and Billings, Montana, have shown that accumulation of emissions from thousands of uncontrolled residential fireplaces and woodstoves can create serious air-quality problems. Because biomass feedstocks are bulky and expensive to transport, facilities for converting them to fuel are also likely to be small and widely dispersed. Currently, small industrial facilities are exempt from many regulations that mandate pollution controls. If this policy continues in the future and is coupled with a large proliferation of small biomass conversion facilities, the result could be poorer air and water quality.

The social and economic ramifications of the use of biomass to produce energy will be diverse and pervasive. The most hotly debated issue may be that of "food versus fuel." Increased demand for biomass feedstocks to produce energy could affect the stability of present markets for many biomass products. The large commitments of land and intensive management required for large-scale production of biomass for energy are likely to cause conflicts with uses of the land for other purposes, such as recreation and wilderness. Other major areas of uncertainty include integration of biomass energy technologies with present farming practice, with rural community patterns, and with existing utility, tax, and regulatory structures.

Biomass energy systems are still in the early stages of development. Consequently, this book represents an evaluation of anticipated environmental and social issues related to exploitation of biomass for energy. However, early evaluations of potential problems do not guarantee that new technologies will develop with minimal environmental and social disruption. Only responsible resource management and planning for mitigation of potential impacts can accomplish such a goal. This book is intended to be a step toward that goal.

2

Existing environment and resource base for biomass-producing systems

Systems for the production and harvest of biomass are a vitally important sector of our economy. These systems currently produce food, fiber, materials, and a small amount of fuel. Biomass-producing systems range from unmanaged ecosystems (e.g., deserts, wilderness forests, oceans) to highly managed systems of agriculture and silviculture. This chapter describes the salient features of existing biomass-producing systems to provide a framework for understanding the implications of biomass energy systems.

Biomass-producing systems can be classified into five major categories: forests, croplands, range and pasturelands, freshwater ecosystems, and marine ecosystems. The U.S. Department of Agriculture (USDA) classified 91% of the land area in the United States as forest, cropland, or pasture (Fig. 2.1) (USDA 1977). Figure 2.2 shows the variation in land use by regions. The use of land for grazing predominates in the Plains and western United States, where 26 to 57% of the land is pasture or range. In the eastern United States, forests occupy 43 to 63% of the total land area, except in the Corn Belt states, where they have been reduced to only 19%. In most regions, cropland accounts for 12 to 36% of land use. However, cropland varies from less than 9% in the mountain states, Alaska, and Hawaii to 55 and 62% in the northern Plains and Corn Belt, respectively.

Significant changes in land use took place between 1967 and 1975 (USDA 1977). About

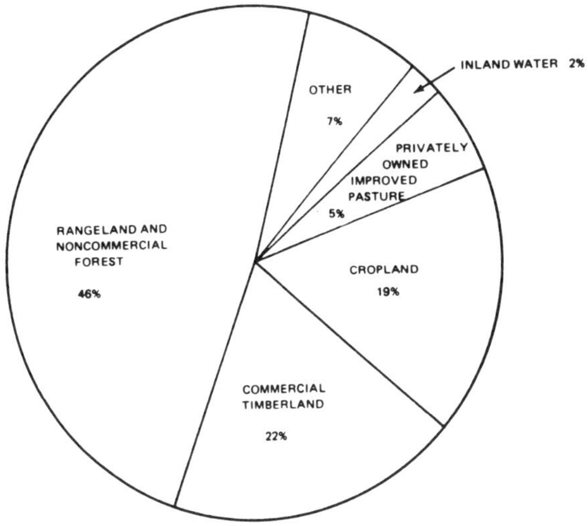

Fig. 2.1. Land use in the United States. Based on total U.S. land area of 936.4 million ha (2313.7 million acres). Source: U.S. Department of Agriculture, 1977. *The Nation's Renewable Resources—An Assessment, 1975*, Forest Resource Report 21, Forest Service, Washington, D.C.

7 million ha (17 million acres) of land were converted to urban and built-up areas, and about 3 million ha (7 million acres) were inundated by water. About 30% of the land converted to urban use came from cropland. Commercial timberland in the West declined as public lands in national forests were reserved as recreation or wilderness areas. In the South much forested land in floodplains and uplands was converted to cropland and

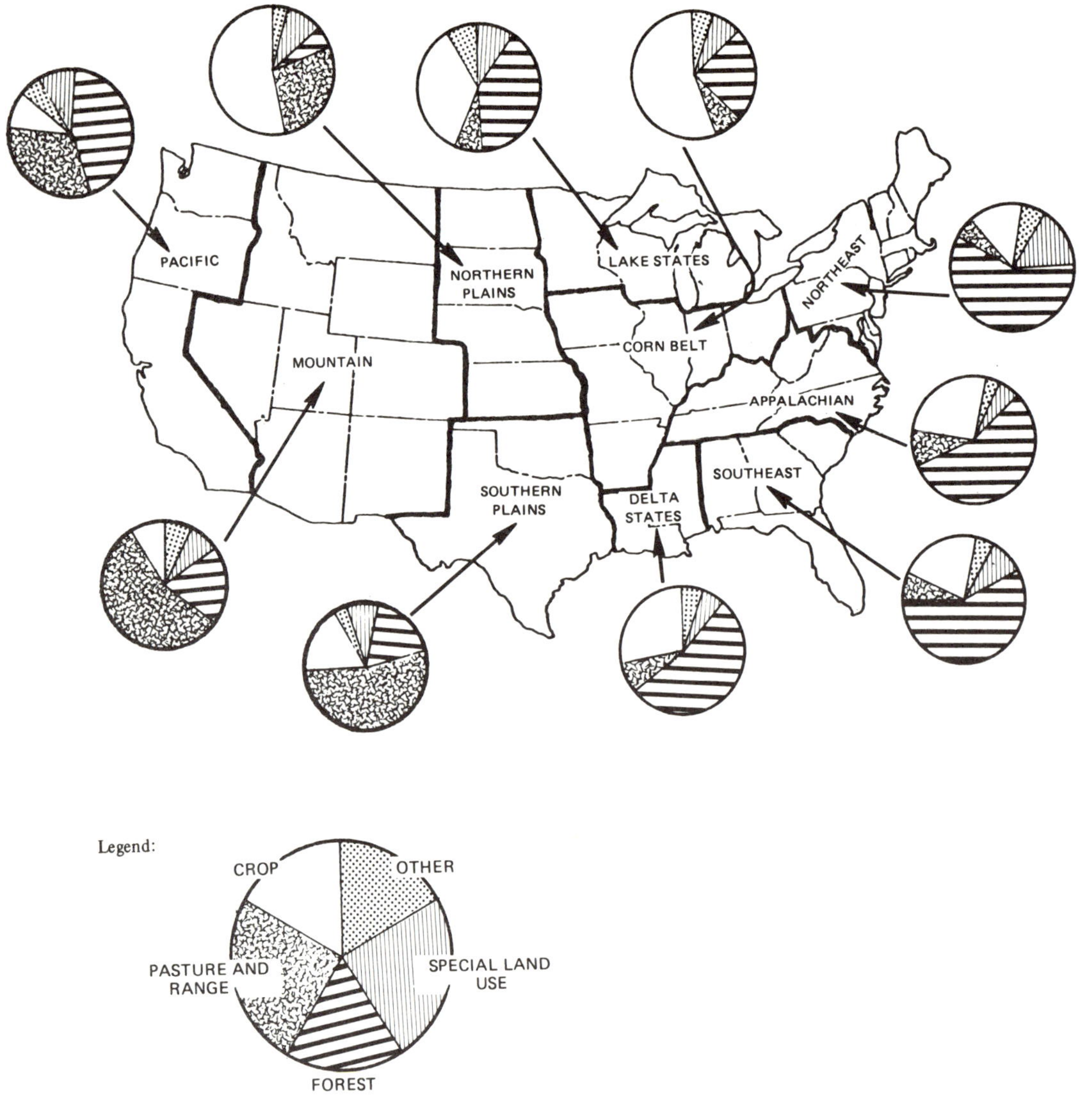

Fig. 2.2. Major land uses by regions in the United States in 1969. Source: U.S. Department of Agriculture, 1973. *Major Uses of Land in the United States—Summary for 1969*, Ag. Econ. Rep.247, Economic Research Service, Washington, D.C.

pasture. Timberland losses were partially balanced by the reversion of abandoned cropland and pasture to forests in the eastern United States. However, the total commercial timberland declined 2% from 1962 to 1970, and this trend continues (USDA 1977). Figure 2.3 summarizes the shifts in land use between 1967 and 1975.

Federal, state, and local governments own about 42% of the U.S. land, with individuals and corporations holding the remainder (USDA 1974). Nearly two-thirds of the federal lands and over one-half of the state-owned lands are located in the West. Most of the federal lands are managed by the Forest Service and the Bureau of Land Management; the rest is administered primarily by the National Park Service, the Fish and Wildlife Service, and the Department of Defense. The proportions of total area owned by major classes of owners, public and private, have been remarkably stable for the past 50 years (USDA 1974).

The existing environments for each of the biomass systems (forests, cropland, grazed land, aquatic systems, and marine systems) are described in greater detail in the following sections.

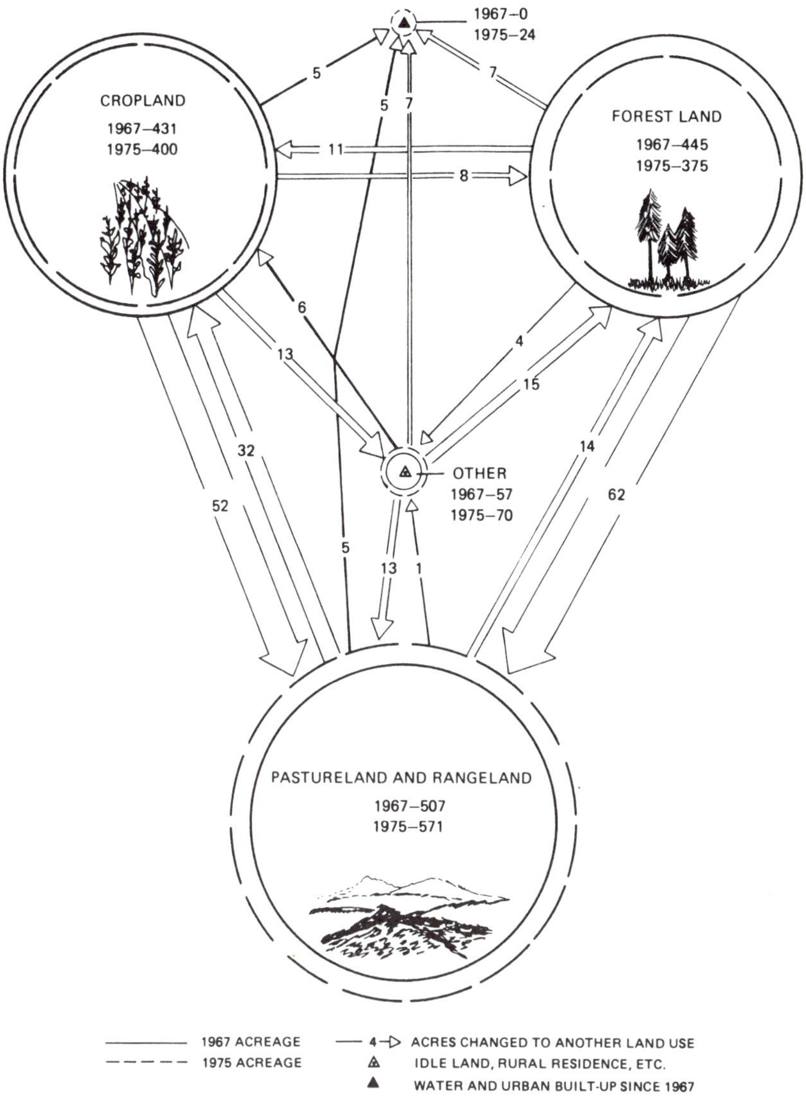

Fig. 2.3. Land-use conversions from 1967 to 1975 (million of acres). Source: Dideriksen, R. I., K. O. Schmude, and A. R. Hidlebaugh, 1977. *Potential Cropland Study*, Soil Conservation Service Statistical Bull. 578, U.S. Department of Agriculture, Washington, D.C.

2.1 Forestry and Silviculture

2.1.1 Land

The land for forestry and silviculturally based biomass consists of existing forests (Table 2.1) and land which could be converted to wood production. There are about 300 million ha (740 million acres) of existing forests in the United States, of which 197 million ha (488 million acres) are classified as "commercial," or able to produce at least 110 kg/ha-year (98 lb/acre-year) (Table 2.2). The inventory of commercial forest land excludes wilderness areas, land reserved for other reasons, and noncommercial land such as pinon-juniper forests which have productivity levels below 110 kg/ha-year. Land that could be converted to wood production exists in semi-arid regions, where shrub production is a possibility, and in regions where land has been taken out of forest for agriculture or some use that does not preclude conversion back to forests or forest plantations. The potential

Table 2.1. Commercial, noncommercial, and other forest land of the United States by regions (millions of hectares)

Category	Total	North[a]	South[a]	Rocky Mountains[a]	Pacific Coast West[a]
Commercial	197.4	69.1	76.3	23.4	28.6
Classes I and II	18.6	3.6	4.7	1.4	8.9
Class III	40.5	13.2	18.4	3.0	5.9
Class IV	83.2	26.2	41.0	6.6	9.4
Class V	55.0	26.1	12.1	12.4	4.4
Productive reserved[b]	7.9	2.1	0.8	3.4	1.7
Productive deferred[c]	1.9	0.1	<0.1	1.3	0.5
Noncommercial	92.4	1.9	6.8	27.7	56.0
Total	299.5	73.2	83.8	55.7	86.8

[a]Forest Service regions defined as follows: north — Maine, N.H., Vt., Mass., R.I., Conn., N.Y., N.J., Pa., Md., Del., W.Va., Ohio, Ky., Ind., Mich., Mo., Iowa, Minn., Kan., Neb., S.D., and N.D.; south — Va., N.C., Tenn., S.C., Ga., Fla., Ala., Miss., Ark., La., Okla., and Tex.; Rocky Mountains — Mont., Idaho, Wyo., Nev., Colo., N.M., and Ariz.; and Pacific Coast — Calif., Ore., Wash., and Alaska.

[b]Public forest land withdrawn from timber utilization through statute or administrative regulation.

[c]National forest lands that meet productivity standards but are under study for possible inclusion in wilderness system.

Source: U.S. Department of Agriculture, 1978. Table 1 in *Forest Statistics of the U.S., 1977* (Draft), GPO Stock No. 001-001-00437-5, Forest Service, Washington, D.C.

Table 2.2. Potential productivity of U.S. forest lands by category

Category	Potential for production of merchantable wood		
	ft^3/acre-year	dry tons/acre-year[a]	kg/ha-year[a]
Class I	>165	>2.5	>909
Class II	120–165	1.8–2.5	661–909
Class III	85–120	1.3–1.8	468–661
Class IV	50–85	0.8–1.3	275–468
Class V	20–50	0.3–0.8	110–275
Noncommercial	<20	<0.3	110

[a]Based on approximate weight of 30 lb/ft^3 of wood.

land resource for silviculture depends greatly on the management technologies that are applied.

The nation's capacity to produce wood depends on the varying patterns of regional land use and productivity. About three-fourths of the nation's commercial timberland is in the eastern United States, whereas noncommercial forests predominate in the West, particularly in the semiarid regions and Alaska. The Pacific Coast has the highest concentrations of both the most productive land and the least productive land (Table 2.1). By comparison, eastern forests (north and south regions) have predominantly intermediate productivity, with two-thirds of the forests producing 110 to 468 kg/ha-year (98 to 416 lb/acre-year).

The federal government holds about 20% of the nation's commercial timberland, most of which is in national forests and two-thirds in the West (Table 2.3). Most of these forests are of

Table 2.3. Ownership patterns of U.S. commercial timberland in 1977 (millions of hectares)

Type of ownership	Total		North	South	Rocky Mountains	Pacific Coast
	Area	Percent				
Federal						
National forest	36.0	18.2	4.1	4.4	14.7	12.7
Bureau of Land Management	2.3	1.2	<0.1	<0.1	0.7	1.7
Other	2.0	1.0	0.4	1.4	<0.1	0.1
Total federal	40.3	20.4	4.5	5.8	15.5	14.5
Indian	2.5	1.3	0.4	0.1	1.1	0.9
State	9.5	4.8	5.3	1.0	0.9	2.3
County and municipal	2.9	1.5	2.4	0.3	<0.1	0.2
Forest industry	27.5	13.9	7.2	14.5	0.8	5.0
Farm	47.4	24.0	18.4	23.2	3.4	2.4
Miscellaneous private	67.3	34.1	30.9	31.4	1.7	3.3
All ownerships	197.4	100.0	69.1	76.3	23.4	28.6

Source: U.S. Department of Agriculture, 1978. *Forest Statistics of the U.S., 1977* (Draft), GPO Stock No. 001-001-00437-5, Forest Service, Washington, D.C.

relatively low site quality and are located at higher elevations, but they nevertheless contain a substantial part of the nation's softwood sawtimber. State, county, and municipal forest holdings are about 6% of the total. The Great Lake states contain 42% of this land, much of which consists of cutover lands that reverted through tax delinquency to public ownership during the depression years of the thirties (USDA 1977).

About 72% of the U.S. commercial timberland is privately owned, with one-third of the total held by individuals and corporations other than the forest industry (Table 2.3). Farmers hold about one-fourth of the total, and forest industries hold only 13.9%. The forest industries, however, hold some of the most productive land. About one-half the forest industry's land is in the South, about one-fourth is in the North, and 18% is in the Pacific Coast region. These holdings put the forest industry in the most favorable position for timber production (Tillman 1978).

Competition for the forest land resource is causing declines in available acreage for possible energy production. For example, forest land may be reserved for parks, wilderness, or other purposes not compatible with production of timber and is consequently designated noncommercial. This land presently amounts to about 8 million ha (20 million acres).

2.1.2 Water

Forest ecosystems have important influences on the hydrologic cycle. Forest vegetation intercepts rainfall and enhances its evaporation. Trees also move large quantities of water from soils into the air by transpiration. However, except in silvicultural nurseries and a few intensely managed plantations, forests are not irrigated. Thus, consumptive use of water is virtually nonexistent in the production and harvest of wood.

In many respects forests play a vital and beneficial role in water management and the hydrologic cycle. Forests decrease storm-water runoff and promote infiltration of precipitation into the soil, recharging groundwater and reducing the intensity of erosion and flooding. In many cases water quality is improved as precipitation passes through forest vegetation and soils. Nutrients, heavy metals, and other potential pollutants are removed, and acidic rainfall may be buffered to a more neutral pH. Municipal water supply systems often own forested watersheds near reservoirs to help maintain reliable supplies of high-quality water (USFS 1975). Although forestry activities sometimes have adverse effects on water quality (Sect. 2.1.5), the existence of large areas of forest in the United States helps to maintain hydrologic stability and water quality.

2.1.3 Energy

Data on energy consumption in forestry are not abundant. The National Academy of Sciences (NAS) Committee on Renewable Resources for Industrial Materials (Bethel 1976*a*) estimated that 3 billion liters (807 million gal) of fuel were used in the harvest of 396 million cubic meters (14 billion ft^3) of timber (1970 level). The other major energy input to forestry is fertilizer, which at the maximum requires roughly the same amount of energy to manufacture as contained in the fuel used by forest-harvesting machinery (Blankenhorn et al. 1978). Thus, a rough estimate of the energy consumed in forestry is 0.25 EJ (0.24 quad), about 10% of that used in agriculture (Sect. 2.2.3). In contrast with agriculture, the production and harvest of timber consume relatively small amounts of energy. However, the resource base of the forest industries is rapidly shifting from old growth to second growth, managed forests, and plantations. This trend will increase the energy demand of timber production. The NAS committee stated that additional data were needed regarding the energy requirements for growing timber (Bethel 1976*b*).

2.1.4 Production and End Uses

The 1977 production from the nation's forests was estimated as about 212 million metric tons (234 million tons) (USDA 1978). This production is expected to increase to 281 million metric tons (310 million tons) by the year 2000 (Tillman 1978). Net forest growth, which is now at about 217 million metric tons (239 million tons), is expected to increase to over 236 million metric tons (260 million tons) by 2000. Under more intensive management this could probably be raised to 308 to 317 million metric tons (340–350 million tons). In 1970 the United States imported approximately 7% of the total wood consumed (USDA 1973).

The primary products from timber are (1) structural wood products, including lumber, plywood, and composition board; and (2) fiber products, including paper, paperboard, market pulp, dissolving pulp, and molded pulp products. Innumerable secondary products are made from these primary products. Minor products from trees are extractives (such as turpentine), chemicals from woody tissue (such as alcohol and phenols), and energy (Jahn and Preston 1976).

Figure 2.4 shows the use distribution of the wood harvested. The forest products industry is undergoing changes in response to a changing raw materials base (Jahn and Preston 1976). There is a diminishing supply of large logs suitable for plywood and large-dimension lumber. There is also an increasing economic incentive to use larger proportions of the forest biomass that was previously left in the forest as residue. These two forces have promoted the production of particle board, flake board, and fiberboard. The industry's resource base is shifting from old growth to managed forests, and the trend toward smaller-sized logs will continue. Thus stiff competition for wood resources already exists. Although this will not preclude use of woody biomass for energy, competition is clearly a constraint to energy development.

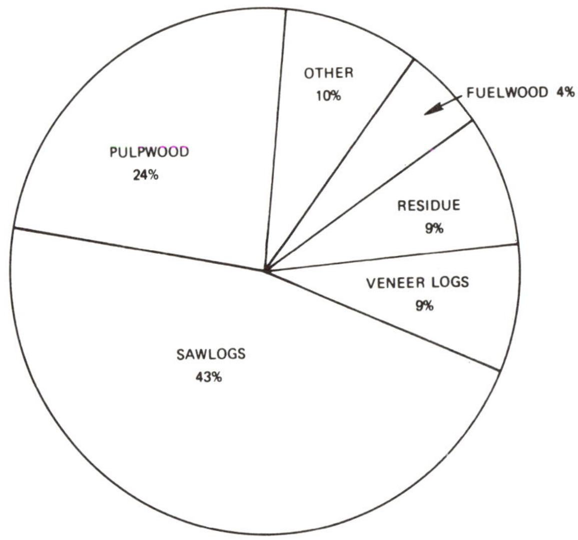

Fig. 2.4. Uses of timber removed from U.S. forests. Percentages add to 99 because of rounding error, based on total removal of 441 million cubic meters (15,555 million ft^3). Source: U.S. Department of Agriculture, Washington, D.C., 1978. *Agricultural Statistics.*

2.1.5 Ecological Aspects

Forest management and harvesting cause a variety of environmental impacts, including changes in hydrology, soils, air and water quality, fish and wildlife, nutrient cycling, and ecosystem structure. These impacts vary greatly, depending on the site and the practices employed. In some cases environmental impacts have been sufficiently detrimental to attract public notice and federal regulation (e.g., the control of nonpoint-source pollution under Sect. 208 of the Clean Water Act). In other cases, however, forest management and harvesting produce inconspicuous or beneficial impacts.

The principal hydrologic effect of forest harvesting is increased intensity and amount of runoff. After logging, peak runoff occurs more quickly after rainfall, sometimes with increased erosion. The impacts of logging vary with both the methods used and the topography; selective cutting produces less hydrologic response than clear-cutting (Douglass and Swank 1975, Kochenderfer and Aubertin 1975). Erosion may occur, not so much from the cutting of trees as from soil disturbance associated with tree removal. Logging roads, skid trails, and landings are most susceptible to erosion. In one study in central Idaho, logging roads caused a 220-fold increase in sediment production, whereas an area with cutting and skidding activity produced only 1.6 times the sediment as an undisturbed area (Megahan 1972).

The practices used in forest management are very important factors affecting environmental impact. The most severe impacts of logging occur when poor forestry practices are used in clear-cutting large tracts. Following clear-cutting, temperatures at the soil surface typically increase, sometimes promoting more rapid decomposition of organic material (Armson 1977). Water quality may decrease for 2 to 20 years after logging because of sedimentation from erosion, increased water temperatures, and release of nutrients from soils. Aquatic systems can suffer especially heavy damage if stream banks are logged or if trees are skidded through streams or along stream banks (Tebo 1955). Burning of logging residues may generate large volumes of particulate emissions, sometimes with adverse effects on air quality.

In establishing new stands following logging, herbicides are sometimes used to control plants that compete with desirable seedlings. Biocides are also used against insect pests and plant pathogens during pest outbreaks. The use of such chemicals over wide areas of forest can affect fish, wildlife, and humans.

A major long-term result of forestry is the replacement of old-growth forest with managed forests or plantations. This practice sometimes involves species transfer, as with the replacement of hardwoods with faster-growing softwoods. The replacement of large areas of mixed-species forests with monocultures and/or even-aged plantations represents a substantial ecological alteration, the large-scale, long-term impacts of which are poorly known.

Forest management also affects wildlife. The removal of cull timber eliminates habitat for some species, such as squirrels and woodpeckers. Clear-cutting favors some wildlife species at the expense of others. For example, deer and grouse prefer open areas created by clear-cuts, whereas species such as the red fox and wood thrush are unable to use early-growth areas.

2.2 Agricultural Crops

Agriculture in the United States today is a highly intensive system of plant, animal, and land management. It is a multibillion-dollar industry that provides food and fiber for both domestic and foreign markets. It is also a primary consumer of resources, making major demands on our land, water, and energy supplies.

2.2.1 Land

Approximately 20% of the nation's total 903 million ha (2.3 billion acres) of land is suitable for the cultivation of crops. This area is now estimated at 162 million ha (400 million acres) [Soil Conservation Service (SCS) 1977] and represents the acreage currently in crop rotation, including land in soil-improvement crops, temporary pasture, or land that is idle (fallow or unused). In 1977, only about 138 million ha (340 million acres) were actually harvested (USDA 1978). Virtually all cropland in the United States is privately owned.

The land area of the 48 conterminous states can be divided roughly into six principal farm production regions (Fig. 2.5) (Thomas and Gilliam 1977).

Fig. 2.5. Farm production regions in the United States.

1. *Corn Belt.* Fertile soils, adequate rainfall, and flat topography combine to make this area the most economical in which to produce corn. The region is also the largest producer of soybeans in the United States. Because corn is raised largely as animal feed, the production of beef cattle and hogs is also important.

2. *Great Plains.* This region resembles the Corn Belt in soils and topography. Lack of adequate rainfall, however, makes production of wheat more feasible. Wheat is the major crop from the Canadian border to Texas. Corn, cotton, and sugar beets are among other crops important to the area. Much or most of the latter three crops are grown under irrigation.

3. *Southeast.* The traditional cotton and tobacco belts are located in this region, although soybeans and corn are today its most important crops. Sugarcane is an important local product in Louisiana and Florida, as are vegetables and fruits in Florida, Virginia, the Carolinas, and Georgia.

4. *Northeast.* The North Atlantic and Great Lake states comprise the dairy belt. Cattle production will be considered further in Sect. 2.3.

5. *Intermountain West.* Solar radiation (insolation) is highest in this area, but lack of rainfall limits the cultivation of crops. Agriculture is dependent on montane water supplies for both irrigation and livestock watering. Farming is of a general crop-livestock nature.

6. *Pacific Coast.* This is the most diverse farming area in the United States. Most of the major crops are cultivated here, as well as several that are unique to the region. The Mediterranean climate and extensive irrigation allow virtually year-round agricultural production in much of the region.

Acreages of cropland in the six farming regions are given in Table 2.4 by capability classes. The capability classification, organized by the Soil Conservation Service (SCS) to reflect levels of suitability of soil to cultivation with sustained yields, is given below (SCS 1966):

Class I. Soils have few limitations that restrict their use.

Class II. Soils have moderate limitations that reduce the choice of plants or require moderate conservation practices.

Class III. Soils have severe limitations that reduce the choice of plants or require special conservation practices, or both.

Class IV. Soils have very severe limitations that reduce the choice of plants or require very careful management, or both.

Table 2.4. Capability classes of U.S. cropland by regions in 1975 (estimated acres)

Region	Class I	Classes II and III	Classes IV—VI	Classes VII and VIII	Total
Northeast	3,100,317	49,949,047	8,086,828	401,512	61,537,704
Corn Belt	10,463,175	69,840,282	6,135,624	290,160	86,729,241
Great Plains	9,928,564	103,783,289	17,968,244	145,216	131,825,313
Southeast	5,414,896	44,107,944	7,035,438	508,397	57,066,675
Intermountain West	1,811,931	28,672,823	9,868,366	279,655	40,632,775
Pacific	2,596,605	14,030,541	5,299,264	0	21,926,410
Total	33,315,488	310,383,926	54,393,764	1,624,940	399,718,118

Source: U.S. Department of Agriculture, 1977. *Potential Cropland Study*, Soil Conservation Service Statistical Bull. 578, Washington, D.C.

Class V. Soils are not likely to erode but have other limitations, impractical to remove, that limit their use largely to pasture, range, woodland, or wildlife habitat.

Class VI. Soils have severe limitations that make them generally unsuited to cultivation and limit their use largely to pasture or range, woodland, or wildlife habitat.

Class VII. Soils have very severe limitations that make them unsuited to cultivation and that restrict their use largely to pasture or range, woodland, or wildlife habitat.

Class VIII. Soils and landforms have limitations that preclude their use for commercial plants and restrict their use to recreation, wildlife habitat, water supply, or to aesthetic purposes.

The numerals indicate progressively greater limitations and narrower choices for practical use. "Prime farmland" is defined as land that has the best combination of physical and chemical characteristics for producing food, feed, forage, fiber, and oilseed crops and is also available for these uses (i.e., is not urban land or water) (*Fed. Regist.* 1978). A farmland inventory based on this definition has not yet been completed, but prime land can be determined by the land capability classification system as including all class I, more than 80% of class II, and less than one-third of class III soils as defined by the SCS (Council on Environmental Quality 1978). Thus, of the 162 million ha (400 million acres) listed in Table 2.4, about 101 million ha (250 million acres) are considered prime lands.

The importance of prime farmland is that it is the most efficient, energy-conserving, environmentally stable land available for meeting food needs. This importance will grow in the future as greater demand for food and fiber must be met while conserving increasingly expensive inputs of energy and materials (Council on Environmental Quality 1978).

2.2.2 Water

Only about 30% of the fresh water provided by annual precipitation is available as runoff or stream discharge. For the conterminous 48 states, this amounts to 4.5 billion m^3/d (1200 billion gal/d) (Water Resources Council 1974). The Water Resources Council (WRC) estimates that of this total, given current surface storage and effects of floods and droughts, only about 25% [1.1 billion m^3/d (278 billion gal/d)] of our renewable surface waters are consistently available for human use. A major portion of runoff is also necessary to sustain the aesthetic, commercial, and recreational values of streams, lakes, and estuaries for man's use as well as to maintain the value of water bodies as habitat for aquatic biota.

Although only 10% of the nation's cropland (17 million ha in 1974) is currently irrigated (USDA 1978), agriculture is the largest consumer of water (Fig. 2.6). The 17 western states account for about 85% of all water consumed in the United States. A severe water

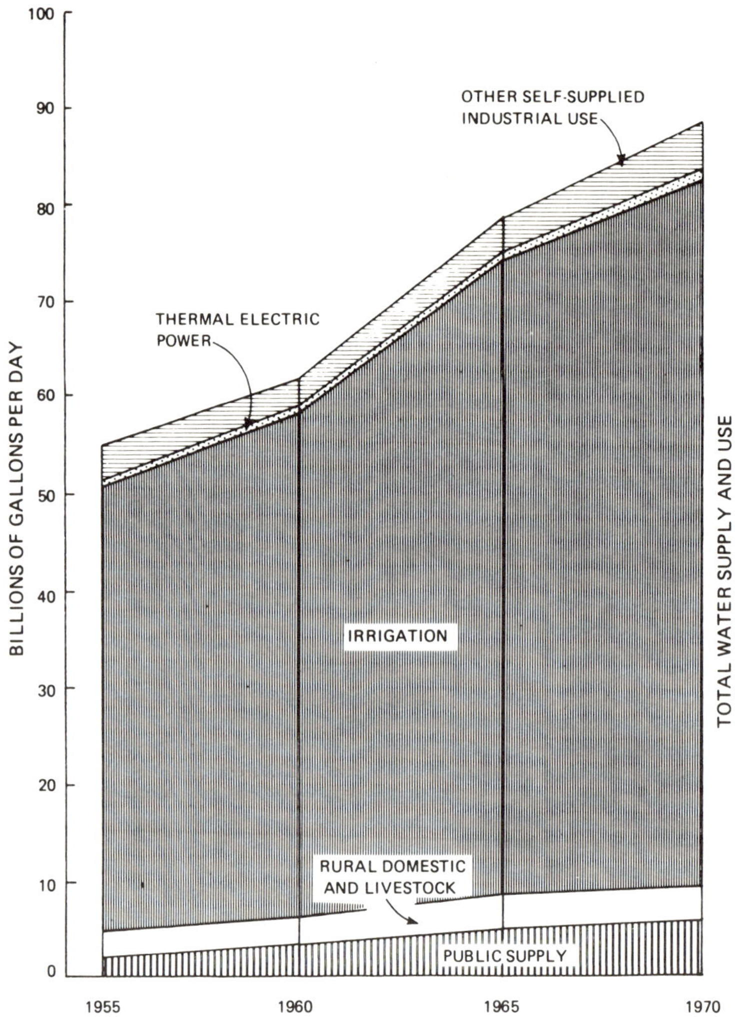

Fig. 2.6. Historic consumption of water for major uses. Source: Water Resources Council, 1974. *Water Requirements, Availabilities, Constraints, and Recommended Federal Actions*, Fed. Energy Admin. Proj. Independence Blueprint, Final Task Force Report, Washington, D.C.

shortage currently exists in 21 WRC subregions, mainly in the central Great Plains and southwestern states (Fig. 2.7). More than 90% of the water in these areas is consumed by agriculture, largely for irrigation (Council on Environmental Quality 1978). A potentially severe shortage is also expected in the northern Plains, Rockies, and northern California by the year 2000 if consumption grows as projected (Council on Environmental Quality 1978) (Fig. 2.7). The western United States is also the area roughly coincident with the greatest amount of insolation (Fig. 2.8), the largest tracts of public land, and the distribution of substantial undeveloped fossil fuel resources in the form of coal and oil shale (Harte and El-Gassier 1978). Vigorous competition for available water already exists there, and any future shifts to industrial or energy development may be expected to intensify this competition.

2.2.3 Energy

Roughly 3% of the U.S. energy budget is accounted for by agriculture (Heichel 1976, Green 1978) (Table 2.5). As an industry, production of raw agricultural commodities ranks third in energy consumption after steel and petroleum refining. About one-half of this energy is consumed directly in the form of petroleum and electricity used in the produc-

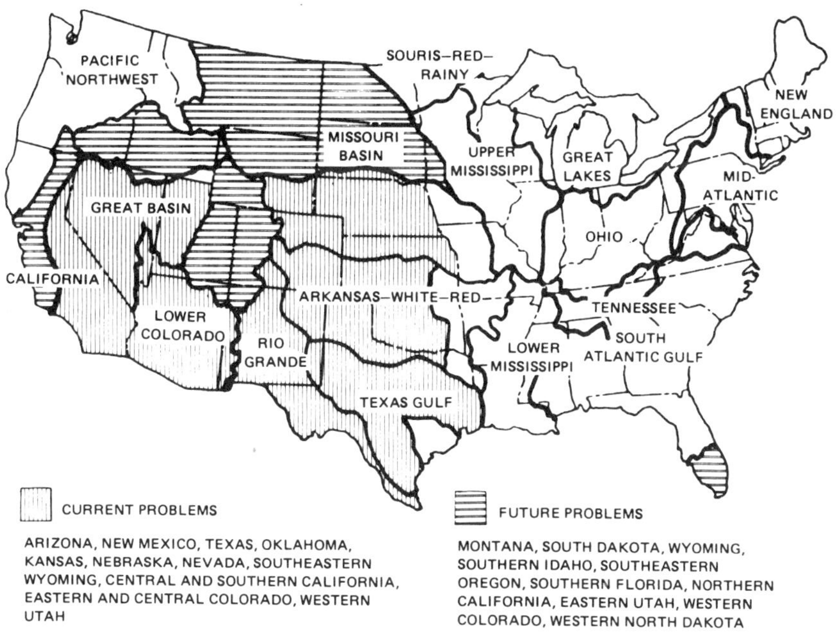

Fig. 2.7. Water supply deficiencies by watershed region. Source: Council on Environmental Quality, Washington, D.C., 1978. *Environmental Quality: The Ninth Annual Report.*

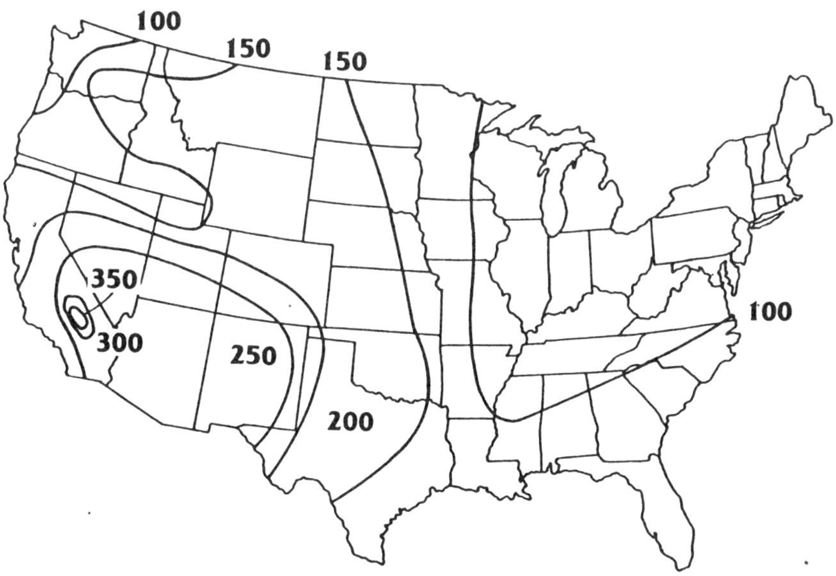

Fig. 2.8. Direct insolation mean daily energy (langleys). Source: McBride, E. J., 1978, *Conference on Environmental Aspects of Nonconventional Energy Resources, II, American Nuclear Society, Denver, Colo., Sept. 26-29, 1978.*

tion of crops and livestock. The remainder consists of purchased inputs: fertilizer, feeds and additives, farm machinery, and pesticides.

Consumption of energy differs greatly for the various types of produce. Field crops such as oats, corn, soybeans, and wheat consume less energy on a per-calorie-output basis than fruit and vegetable cultivation. The ratio of energy input, or "cultural energy," to output of food energy is one measure of energy efficiency (Heichel 1976). Figure 2.9 is a graph of efficiencies for 24 grain, forage, fruit, and vege-

Table 2.5. **Energy consumed annually by agriculture in the United States (1974 data)**

	Energy (EJ)[a]	Agricultural use (%)
Farm production		
Crops		
Petroleum	0.79	38
Electricity	0.01	<1
Livestock		
Petroleum	0.26	12
Electricity	0.04	2
Purchased inputs		
Fertilizer	0.59	28
Petroleum	0.21	10
Feeds and additives	0.11	5
Animal and marine oils	0.04	2
Farm machinery	0.03	1
Pesticides	0.01	<1
Total	2.09	

[a] 1 EJ = 10^{18} J ≈ 1 quad.

Based on data from Economic Research Service, U.S. Department of Agriculture, 1974. *The U.S. Food and Fiber Sector: Energy Use and Outlook*, prepared for the Committee on Agriculture and Forestry, U.S. Senate, Washington, D.C.

table crops. The largest acreages in the United States are devoted to the most efficient crops (corn, wheat, soybeans, and sorghum). These crops are nevertheless expensive in terms of energy input because of the vast areas over which they are cultivated. Corn, for example, is the biggest single user of fertilizer in the United States. This crop is produced on about 25 million ha and accounts for one-third of the nitrogen, 36% of the phosphorus, and 43% of the potassium used in the United States for fertilizer (Thomas and Gilliam 1977). Percentages of fertilizer applied to other crops are lower and vary with the needs of particular plants.

Overall, the marked increase in yields of agricultural products over the past 30 to 35 years is largely attributable to an equivalent increase in input of fossil fuels (Pimentel et al. 1973). For example, from 1945 to 1970, mean corn yields increased from about 34 to 81 bushels/acre while the associated mean energy inputs increased from 3.8×10^9 to 1.2×10^{10} J. Hence, the energy efficiency of corn production has decreased, although yields

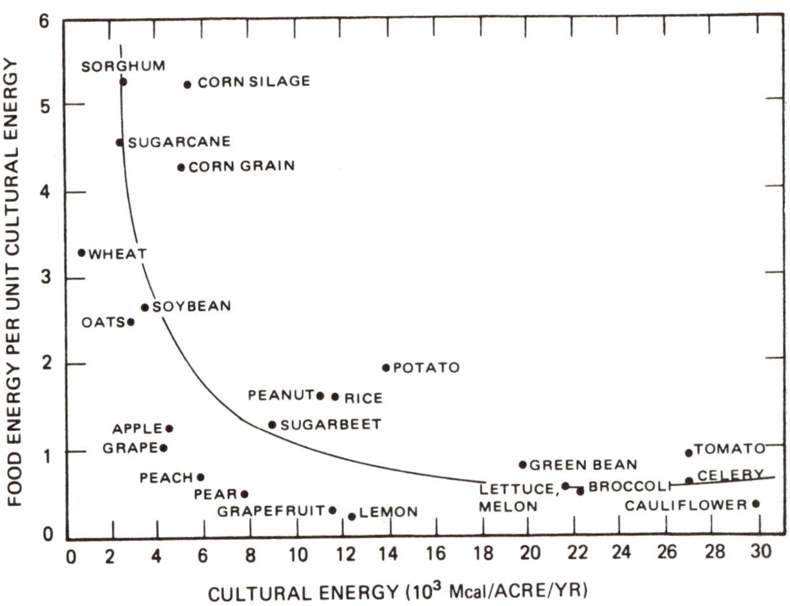

Fig. 2.9. Cultural energy used to produce 24 grain, forage, fruit, and vegetable crops and the amount of food energy (in calories) that each plant yields per unit of cultural energy invested. The ratio is a measure of food energy efficiency. Source: Heichel, G. H. A., 1976. "Agricultural Production and Energy Resources," *Am. Sci.* **64:** 64-72.

have increased. Increase in farm output as a function of energy input may have reached, or may soon reach, the point of diminishing returns (Steinhart and Steinhart 1974, Jensen 1978). This point of view is evident in the apparent leveling off of the curve in Fig. 2.10. American agriculture is now heavily dependent on a regular input of energy in the form of fossil fuels. If further increases in productivity are possible, they may be expected to constitute an even greater energy expense unless alternative means of achieving such increases are found.

2.2.4 Production and End Uses

Table 2.6 summarizes production of the major agricultural crops for 1976. Acreages harvested and dollar values are given to provide relative magnitudes. The USDA also

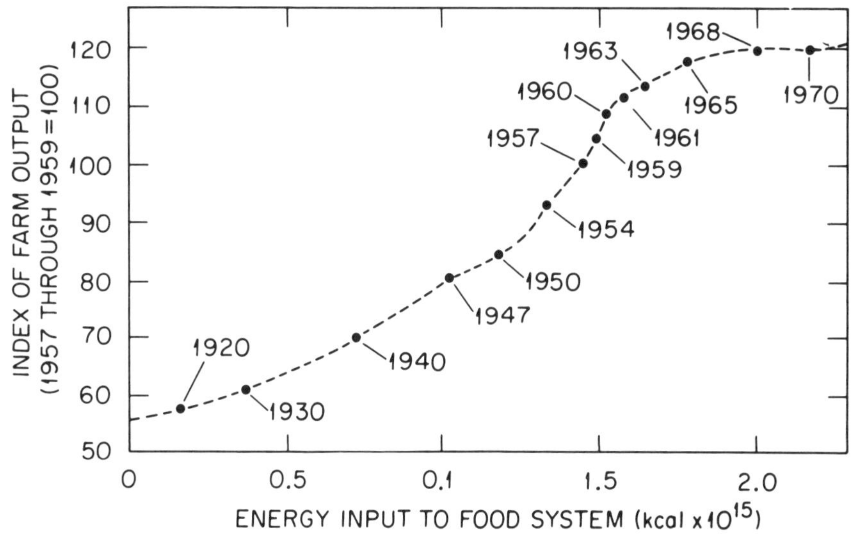

Fig. 2.10. Farm output as a function of energy input to the U.S. food system, 1920 through 1970. Source: Steinhart, J. S., and C. E. Steinhart, 1974. "Energy Use in the U.S. Food System," *Science* **184:** 307-16.

Table 2.6. Land use and production of major agricultural crops in 1976

Crop	Area harvested (thousands of acres)	Production (thousands)	Value of production (thousands of dollars)
Corn grain	71,300	6,266,359 bu	13,471,796
Corn grain silage and forage	12,130	117,813 ton	
Wheat	70,771	2,142,362 bu	5,851,443
Oats	11,946	546,315 bu	845,188
Barley	8,297	372,461 bu	829,716
Rice	2,480	115,648 cwt	811,358
Sorghum grain	14,723	719,817 bu	1,450,085
Sorghum grain forage and silage	2,655	7,168 ton	
Hay	60,311	120,006 ton	6,810,799
Soybeans	49,358	1,287,560 bu	8,768,979
Peanuts	1,522	3,750,890 lb	750,260
Sugarcane	747	28,120 ton	257,242
Sugar beets	1,478	29,386 ton	616,813
Vegetables	3,296	Variable	3,120,261

Source: U.S. Department of Agriculture, Washington, D.C., 1978. *Agricultural Statistics.*

provides statistics on domestic versus export disposition of commodities. From 1972 through 1977, about 30% of the area harvested annually produced crops for export worth roughly $24 billion in 1977 (USDA 1978). An additional 60% of the acreage for field crops produces livestock feeds, whereas only 10% is devoted to production for other domestic (i.e., human) use.

In addition to food and fiber for consumption, crop harvesting also produces large quantities of associated residues. The amounts and current uses of these materials have been estimated to vary between 251 and 354 million dry metric tons (Burwell 1978, Klass 1976, Stanford Research Institute 1976). Table 2.7 summarizes Stanford Research Institute's survey estimates of quantities and current disposition of agricultural residues. Roughly 75% of these materials are returned to the soil, mostly at the site of production. Only 2% are considered wastes—that is, materials that must be disposed of at a cost. The principal method of disposal is burning. (Over 90% of the materials in this category are cotton-gin trash and rice, sugarcane, and seed-grass residues.) The remainder is collected and sold (4%), used as fuel without sale (1%, chiefly in southeastern states), or fed to livestock (19%) with or without collection but without sale, especially in regions where livestock density is high. The nature of residues available in any given region will obviously depend on the predominant crops grown. Table 2.8 lists examples of residues and regions of greatest production. Corn, wheat, and soybeans together account for over three-quarters of the residues produced (Larson, Holt, and Carlson 1978). The energy content of these materials is of great interest in the consideration of biomass fuels and feedstocks.

Table 2.7. Disposition of agricultural residues in the conterminous United States

Disposition	Amount (tons dry weight)	Percent
Sold[a]	11,266,626	4
Fed[b]	52,142,254	19
Fuel[c]	1,741,990	1
Returned[d]	205,599,417	74
Wasted[e]	6,810,630	2

[a]Available residues collected and sold for purposes other than fuel.

[b]Available residues fed to animals, with or without collection but without sale.

[c]Available residues used as fuel, but without sale.

[d]Available residues returned to the soil, but without sale.

[e]Available residues that must be disposed of at an economic cost; for example, residue removed from the land for disposal or burned in place. This category includes all collected crop residues not sold, fed, or used as fuel and involves immediate or subsequent costs for returning them to the soil.

Source: Stanford Research Institute, Menlo Park, Calif., 1976. *Crop, Forestry, and Manure Residue Inventory,* prepared for U.S. Energy Research and Development Administration.

Table 2.8. Annual production of residues for various U.S. crops and regions of highest production

Crop	Total U.S. residue produced (metric tons × 10^3)	Areas of highest production[a]
Corn grain	187,282	Corn Belt, Great Plains
Wheat	81,830	Great Plains, northwest
Soybeans	35,500	Corn Belt, southeast
Sorghum grain	12,080	Great Plains
Oats	25,882	Northeast, Great Plains
Rye	30,291	Great Plains
Barley	16,042	Great Plains, Pacific
Rice	6,635	Southeast
Peanuts	2,074	Southeast

[a]Areas of highest production were derived from a partial list of states given in Midwest Research Institute, Kansas City, Mo., 1978. *Systems Study of Fuels from Grains and Grasses, Final Report,* Prepared for U.S. Department of Energy.

2.2.5 Ecological Aspects

The agricultural system is not without its intrinsic environmental effects, both on its own resource base and on the natural environment. Several issues are of such magnitude as to deserve mention before considering intensification or addition of factors related specifically to production of biomass for fuels.

2.2.5.1 Loss of productive land. Erosion of soil by water results in delivery of about 3 to 4 billion tons of sediment to the nation's waterways each year. About three-fourths of this comes from agricultural land (Council on Environmental Quality 1978). Wind erosion accounts for the loss of another billion tons of agricultural soil annually. Dust storms similar to those of the 1930s recurred on the Great Plains in 1976 and 1977 with erosion damage affecting an acreage comparable to that of the worst year (1938–1939) of the Dust Bowl (Lockeretz 1978). Unfortunately, serious erosion losses are not now confined to the Dust Bowl region. More than one-third of all U.S. cropland is estimated to be suffering soil loss too great to be sustained without an ultimately disastrous decline in productivity (Carter 1977), and two-thirds of all crop and pasture land is in need of additional soil protection measures (Council on Environmental Quality 1978). The threat of loss of productivity is real. Since 1935, about 40 million ha (100 million acres) of land have been degraded beyond the point of sustaining further cultivation. Another 40 million ha have lost more than 50% of their original topsoil (Council on Environmental Quality 1978). Soil conservation measures adequate to sustain long-term agricultural productivity have not yet been achieved despite the expenditure of nearly $15 billion since the 1930s.

The causes of this situation are both political and economic. Contradictory federal policies have encouraged marginal land set-aside and other conservation measures on the one hand while calling for all-out crop production on the other. Efforts to maximize short-term crop yields in response to large increases in commodity prices, particularly for exported food, have taken precedence over longer-term efforts to save the soil. Practices such as abandonment of crop rotation for continuous planting of wheat, removal of shelter belts, and plowing and use of highly erodible marginal land have been identified as destructive to soil conservation efforts. The severity of the erosion problem has prompted recent proposals for legislative action to enforce stricter conservation practices (Carter 1977).

In addition to cropland loss caused by agricultural abuse, about 30% of the land converted to urban, highway, and reservoir use each year comes from cropland. Prime farmland, in particular, is being withdrawn from food and fiber production at a disproportionate rate amounting to some 1 million acres/year, or somewhat over 4 sq miles/d (Council on Environmental Quality 1978). Losses are most acute in the Pacific and northeast regions, but the Appalachians and Great Lakes are also under heavy pressure.

To make up for loss of cropland, land now in pasture and range will have to be converted. The SCS has recently had to revise its estimate of land available for conversion from the 1967 figure of 107 million ha (265 million acres) to only 45 million ha (111 million acres) (SCS 1977). Only 14 million ha (34.9 million acres) of this land has high potential for conversion to cropland without requiring installation of costly conservation measures. Thus, the United States has a declining active agricultural land base, its cropland reserves are questionable, and the economic feasibility of future conversions is uncertain (Carter 1977; Doering 1977). Disagreement has arisen over how soon and how severely our cropland resource base will be tested. Some analysts believe this could occur as early as 1985 (Schiff 1979). A recent USDA study (Lee 1978) suggested that a combination of increased export demand; constraints on use of water, energy, and agricultural chemicals; and unfavorable weather conditions could result in a tremendous need for additional cropland in the near future. An economic projection model incorporating only one of these factors—increased export demand—projected a 33 million ha (81 million acre) increase in harvested cropland by 1985.

2.2.5.2 Agricultural runoff. Runoff of soils, fertilizers, and pesticides from farmland and irrigation return flows has emerged as among the most significant sources of water pollution in the nation (Train 1978). Agriculture is the most extensive source of nonpoint-source pollution, affecting over half of the water basins in all geographic regions of the United States (Council on Environmental Quality 1978). About 3 billion metric tons of water-borne sediment from agricultural land is settling annually in reservoirs, rivers, and lakes, shortening the usable lifespan of rural watersheds and municipal water supplies and smothering aquatic life. Sediment carries with it fertilizers and pesticides that load water courses with toxicants and high biological oxygen demand (BOD) materials which deplete oxygen levels and make water uninhabitable.

If not curbed, nonpoint-source pollution will prevent attainment of the national water quality goals mandated by the Clean Water Act. To date, progress has been minimal (Council on Environmental Quality 1978). Runoff from farmland is precipitation-dependent; thus, inputs are erratic. The best method of control is not water treatment but management of land (Frere et al. 1977).

2.3 Livestock Production

The per-capita consumption of meat in the United States in 1976 was about 101 kg (222 lb) (USDA 1978). In addition, Americans consumed about 176 kg (388 lb) of eggs and dairy products per person. The produce from about 60% of our field cropland, or about 90% of the cereal, legume, and vegetable protein suitable for human use, is fed to livestock to convert vegetable protein to high-quality animal protein (Pimentel et al. 1975). The raising and maintenance of livestock is thus a predominant aspect of our agricultural system, both in the products it supplies and in the resources it uses.

2.3.1 Land

The USDA estimated that livestock grazed on 360 million ha (890 million acres), or 39% of the land area of the 50 states in 1969 (USDA 1974). About two-thirds of this 360 million ha was on farms. Of the nonfarm grazing land, about one-half was federal grassland, one-fourth was federal woodland, and the remainder was chiefly woodland in state or private ownership. Most of the federal land grazed is in the semiarid western states and has low productivity.

The greatest beef cattle and hog production is in the Corn Belt and the Great Plains. Increasing specialization and mechanization are resulting in large-scale removal of beef cattle and hogs from pasture and grassland and greater production in confinement. Large commercial feedlots are now abundant in the high plains areas of Texas, Oklahoma, Kansas, Nebraska, and Colorado. The overall trend is toward an increase in the number of feedlots and in the number of lots with large capacity (Table 2.9) (Loehr 1974).

Table 2.9. Beef cattle feedlot numbers, capacities, and marketing

Capacity (number of head)	Number of feedlots			
	1962	1964	1968	1970
1,000—1,999	752	808	967	1004
2,000—3,999	373	421	522	549
4,000—7,999	179	242	316	333
8,000—15,999	105	120	176	210
16,000—31,999	26	34	80	105
32,000+	5	10	19	41
Feedlot with capacity of 1000 or more cattle				
Percentage total feedlots	0.64	0.75	0.99	1.1
Percentage beef cattle marketed		39	47	55

Reprinted by permission of Loehr, R. C., 1974. *Agricultural Waste Management — Problems, Processes, and Approaches,* Academic Press, New York.

Dairy farming is heavily concentrated in the North Atlantic and Great Lake states. The number of dairy cows nationwide decreased from 21 million head in 1955 to 13 million head in 1968, but average milk production per cow increased sufficiently to maintain dairy production at an almost constant level. Small dairy farms (those with fewer than 30 cows) are disappearing, whereas large farms are on the increase. As in beef cattle production, the trend is toward increasing production in confinement. This trend has altered the traditional complementary relationship between crop and livestock production: the feed produced on the land served for livestock and the manure from the livestock went back to the land. Large confinement operations increasingly import feed from adjacent farms and states. Handling and disposing of livestock wastes has, in turn, become an environmental quality problem (Loehr 1974) (Sects. 2.3.4 and 2.3.5).

2.3.2 Water

In 1965 about 0.43 billion cubic meter (112.6 billion gal) of water was withdrawn and 0.25 billion cubic meter (66.3 billion gal) was consumed (i.e., not returned) per day by agriculture. Only 2% of each category was devoted to livestock; the remainder went for irrigation. Thus, the water demand associated with livestock production is small compared with crop production needs. Raising livestock does, however, have a considerable impact on water quality. This impact will be discussed further in Sect. 2.3.5.

2.3.3 Energy

The introduction of an intermediate consumer, such as a cow, into a food chain entails an order-of-magnitude reduction of energy efficiency; that is, the caloric energy available by consuming meat fed on grain is only about 10% of that directly available in the grain (Heichel 1976). This loss of energy is largely due to the metabolic needs of the food animals.

In terms of fossil fuel consumption, livestock fed only grains grown on croplands requires a considerable energy subsidy that amounts to about 10 calories for each calorie of food output (Steinhart and Steinhart 1974). Range-fed animals, however, can feed off of land that is unsuitable for the cultivation of crops and therefore require little energy subsidy. Hence, this is a more efficient means of providing human food calories and protein. Unfortunately, range-fed livestock alone provides meat in quantities insufficient to meet the demands of American consumers. Consequently, the majority of our agricultural land is devoted to the production of animal feedstuffs (Sect. 2.2.4).

2.3.4 Production and End Uses

In addition to the food and fiber resulting from livestock farming, large quantities of animal wastes (manures) are produced. Table 2.10 summarizes one estimate of the number of farm animals and associated wastes produced in the United States. Approximately 88% of total manure production is derived from cattle,

Table 2.10. U.S. livestock animals and associated wastes produced annually (1971 data)

Animals	Number of animals ($\times 10^3$)	Amount of solid organic waste generated[a]	
		Each animal/day [kg (lb)]	All animals/year [metric tons (tons) $\times 10^6$]
Cattle	114,568	3.72 (8.20)	156 (172)
Hogs	67,540	0.41 (0.91)	10.2 (11.3)
Sheep	19,560	0.21 (0.46)	1.49 (1.64)
Poultry	1,116,680	0.02 (0.047)	8.65 (9.54)
Total			176.4 (194.5)

[a]Moisture and ash free.

Source: Environmental Protection Agency, Washington, D.C., 1978. *Preliminary Environmental Assessment of Biomass Conversion to Synthetic Fuels.*

6% from swine, 1% from sheep, and the remainder from poultry (USEPA 1978). Roughly one-half of these wastes is produced by animals in feedlots and confined rearing situations, with this amount increasing because of trends in the feedlot industry. The remaining one-half is produced by animals in pasture and cropland where collection is not feasible. Estimates of the amounts of dry-weight manure available for energy production vary from 24 million metric tons (Stanford Research Institute 1976) to as much as 218 million metric tons (Stephens and Heichel 1975). The results of a Stanford Research Institute (SRI) inventory (Table 2.11) indicate that 67% of collectable manure is currently returned to the land, whereas 15% is sold, presumably as fertilizer, and 17% must be disposed of as waste.

The type of manure disposal employed in a feeding operation depends on the type of farm animal and the location of the facility (USEPA 1978). Liquid-disposal systems are common in dairy operations and in colder and damper regions. Liquid manure is pumped through a sewer to a containment pond (ditch, septic tank, or lagoon), where bacterial decomposition occurs over a long period of time. Solids-disposal systems are used extensively on large beef feedlots. Manure deposited on the open ground is periodically scooped into mounds, where composting processes occur rapidly. Several times a year, this material is loaded onto trucks and hauled to local farms for use as fertilizer. Large feedlots must also have lagoons to accommodate runoff during wet weather. In areas where large feedlots have become common (e.g., Colorado, Texas, Arizona, and California), manure supply has exceeded demand in recent years so that it remains stockpiled on feedlots (USEPA 1978).

Table 2.11. Quantities and disposition of manure available for conversion to energy in the conterminous United States

Disposition	Quantity (tons dry weight)	Percentage
Sold[a]	3,920,283	15
Fed[b]	247,064	1
Fuel[c]	16,436	>1
Returned[d]	17,845,698	67
Wasted[e]	4,429,867	17
Total	26,459,348	

[a]Available manure collected and sold for purposes other than fuel.

[b]Available manure converted to animal feed, with or without collection but without sale.

[c]Available manure used as fuel, but without sale.

[d]Available manure returned to the soil, but without sale.

[e]Available manure that must be disposed of at an economic cost. This category involves immediate or subsequent costs for return to the soil.

Source: Stanford Research Institute, Menlo Park, Calif., 1976. *Crop, Forestry, and Manure Residue Inventory*. Prepared for U.S. Energy Research and Development Administration.

2.3.5 Ecological Aspects

The major environmental concern about livestock raising is the disposal of waste materials. Concentrated animal feeding operations are significant sources of water pollution in some regions. Storm-water flushing of animal wastes from feedlots or runoff from disposal lagoons and/or manured fields may affect the quality of adjacent streams. Fish kills and deterioration of recreational areas have resulted (Loehr 1974). Table 2.12 gives ranges of values for runoff from nonpoint sources of water pollution. Animal feedlot runoff and manure seepage characteristically yield pollutant quantities several orders of magnitude higher than those of other nonpoint sources.

2.4 Freshwater Systems

2.4.1 Description

The biomass produced in freshwater has a small role in our economy compared with that produced on farms, forests, and rangeland. Furthermore, rivers, lakes, ponds, reservoirs, and waterways have diverse uses, which include water supply, recreation, navigation, flood control, and power generation. Most aquatic systems are not managed specifically for biomass production. Thus, except for sport-fish production, biomass production in fresh water is rather inconspicuous compared with that of agriculture or forestry.

The U.S. water bodies vary widely in size, shape, and use. The spectrum includes ponds,

Table 2.12. Summary of nonpoint source characteristics[a]

Source	Concentration (mg/liter)					Annual area yield rate (kg/ha)					Surface area of interest
	COD[b]	BOD[c]	NO_3-N	Total N	Total P	COD	BOD	NO_3-N	Total N	Total P	
Precipitation	9-16	12-13	0.14-1.1	1.2-1.3	0.02-0.04	124		1.5-4.1	5.6-10	0.05-0.06	Total land area
Forested land			0.1-1.3	0.3-1.8	0.01-0.11			0.7-8.8	3-13	0.03-0.9	Forest area
Rangeland								0.7		0.08	Rangeland
Agricultural cropland	80	7	0.4	9	0.02-1.7				0.1-13	0.06-2.9	Active cropland
Land receiving manure									4-13	0.8-2.9	Crop or unused land used for manure disposal
Irrigation tile drainage, western United States											
Surface flow			0.4-1.5	0.6-2.2	0.2-0.4				3-27	1.0-4.4	Irrigated western soils
Subsurface drainage			1.8-19	2.1-19	0.1-0.3			83	42-186	3-10	Irrigated western soils
Cropland tile drainage				10-25	0.02-0.7				0.3-13	0.01-0.3	Active cropland requiring drainage
Urban land drainage	85-110	12-160		3	0.2-1.1	220-310	30-50		7-9	1.1-5.6	Urban land areas
Seepage from stacked manure	25,900-31,500	10,300-13,800		1,800-2,350	190-280						Manure holding area
Feedlot runoff	3,100-41,000	1,000-11,000	10-23	920-2,100	290-360	7,200	1,560		100-1,600	10-620	Confined, unenclosed animal holding areas

[a] Data do not reflect the extreme ranges caused by improper waste management or extreme storm conditions; the data represent the range of average values reported in previous tables.
[b] Chemical oxygen demand.
[c] Biological oxygen demand.
Source: Environmental Protection Agency, 1978. *Preliminary Environmental Assessment of Biomass Conversion to Synthetic Fuels*, Washington, D.C.

lakes, reservoirs, streams, rivers, canals, waterways, wetlands, and highly managed systems such as sewage lagoons, cooling ponds, and aquaculture systems. The Forest Service (USDA 1977) classified 19 million ha (48 million acres) as inland water. This amount includes water bodies larger than 16 ha (40 acres) and streams greater than 0.2 km (0.25 mile) wide; the Great Lakes are excluded. Nineteen million hectares represents only 2.1% of the U.S. land area. States with greater than 3% of their surface area designated as inland water include Wisconsin, Minnesota, Alaska, and many Atlantic and Gulf Coast states, especially Louisiana, Florida, North Carolina, Delaware, Rhode Island, and Maine. In addition, the Forest Service classifies 1.8 million ha (4.5 million acres) as wet grasslands, occurring primarily along the Atlantic and Gulf coasts (e.g., Florida's Everglades) and in Texas and Louisiana (USDA 1973). Finally, the USDA's 1967 Conservation Needs Inventory identified 107 million ha (265 million acres) of cropland as having excess water sufficient to limit regular cultivation. Much of this land is located in the delta and northern lake states regions and in the southeastern coastal area (USDA 1974).

Because inland waters are so varied, information describing them must be drawn from many sources. No single government agency is devoted to inland waters; instead, responsibility for various aspects (navigation, water quality, etc.) is spread throughout many agencies. In an effort to give an overview of freshwater systems, brief descriptions of various types of inland water are provided:

1. *Glacial lakes.* These lakes occur throughout the glaciated areas, which include the northern United States and high Rocky Mountains. Water quality is often excellent, although human activities have sometimes caused nutrient enrichment and nuisance algal growth. Most natural lakes are of glacial origin. The Great Lakes are good examples of large, glacially derived water bodies.

2. *Rivers and impoundments.* Most of the major river systems in the United States are dominated by impoundments that create two types of reservoirs: main stream reservoirs, used for navigation and flow regulation, and storage reservoirs. Both types typically serve for recreation, water supply, and power generation. Nuisance growth of rooted aquatic plants sometimes occurs in eutrophic reservoirs, although this is more common in lakes because of steady water levels.

3. *Farm ponds.* The SCS encouraged the construction of many ponds following the 1930s. These ponds provide erosion control, stock watering, limited irrigation, fishing, swimming, and boating. Perhaps one million farm ponds are distributed throughout the rural United States (Bennett 1970). Shallow areas of farm ponds may become choked with aquatic vegetation, such as cattails or bulrushes. In the South many ponds are specifically built for fish farming.

4. *Excavated ponds and lakes.* These ponds and lakes may be created in quarries, gravel pits, strip-mine depressions, or excavations specifically to form a water body. Water quality is highly variable and may limit productivity.

5. *Swamps and marshes of the coastal plain.* Low-lying areas of the coastal plain have large areas of wetlands, such as the Great Dismal Swamp, Okefenokee Swamp, and the Everglades. These areas may be dominated by cypress forest or swamp prairie (wet grassland). Such large swamps and marshes are typically interconnected with the water table, and large-scale alterations of groundwater can result when canals are built to drain these wetlands. These wetlands are also important habitat for wildlife. Estuarine and brackish marshes also occur generally along the Atlantic and Gulf coasts and especially in the Mississippi River delta.

6. *Midwestern potholes and wetlands.* Throughout those parts of the Midwest that were glaciated, numerous potholes, sloughs, and bogs occur, which are remnants of glacial lakes that have filled in. Both hard- and soft-water conditions are found. These areas are important to migrating waterfowl.

7. *Oxbow and prairie lakes.* In the unglaciated parts of the Midwest, natural lakes are not abundant. Along large rivers, however, oxbow lakes have formed from isolated river meanders. In the plains region depressions occur that retain water during wet weather. Because of rapid evaporation, total dissolved solids increase, sometimes to very high levels.

8. *Water bodies of arid lands.* The low rates of precipitation and high rates of evaporation cause much water in the West to have high concentrations of dissolved solids. Saline lakes commonly occur in southern California and the Great Basin. This natural condition has been intensified by large-scale irrigation and diversion of water. Many impoundments exist on river systems of the arid West and the Pacific Coast.

2.4.2 Harvest and End Uses

2.4.2.1 Fish production. The National Marine Fisheries Service reported that commercial fishing produced 57,300 metric tons (154 million lb) of freshwater fish in 1974 in the Great Lakes and Mississippi River system fisheries (National Oceanic and Atmospheric Agency 1977). The primary species were alewives, buffalofish, and carp. As shown in Table 2.13, fish farms produced about 13,400 metric tons of salmon and trout in 1972 and 38,000 metric tons of catfish in 1975 (Walsh 1978).

Aquaculture of salmon is conducted along the Pacific Coast primarily by using ocean ranching, in which fish are released to the ocean as juveniles and harvested as maturing adults. Trout are grown where cold water is most abundant, principally in Idaho and Montana, and are reared to market size in raceways or ponds with rapid flow-through. Both trout and salmon require high-quality, protein-rich food, and the high costs of aquaculture make those fish a luxury food.

Channel catfish are raised in fish ponds in the southern states, with 80% of the total coming from Mississippi, Louisiana, and Arkansas (Walsh 1978). Ponds range in size from 0.1 to 10 ha and larger, with 0.4-ha (1-acre) ponds being quite typical (Brown 1977).

2.4.2.2 Aquatic plant production. Throughout the South and other parts of the United States, the growth of rooted aquatic plants such as the water hyacinth often reaches the point of nuisance, especially when the water body is enriched by nutrients from wastewater or nonpoint sources. Water hyacinth causes particularly serious problems by obstructing navigation, reducing water flow, and hindering recreation. Decomposition of aquatic plants also reduces water quality by consuming dissolved oxygen, sometimes causing fish kills. In Florida, Louisiana, and Texas, 700,000 ha are infested with water hyacinth, and these three states alone spend over $11 million per year for control (Wolverton and McDonald 1979). The primary control of aquatic nuisance growth is by herbicides such as 2,4-D.

Mechanical removal of water weeds, another control method, has the advantage of removing some of the nutrients from the aquatic system. However, mechanical harvest is expensive, handling and transportation of the harvested material are difficult, and a large-scale economic use for the harvested biomass does not yet exist. Many uses appear feasible, however, including use as livestock feed, fiber for materials, soil additives, and feedstock for alcohol or methane production (Del Fosse 1977). Because no estimates could be found of the quantity of waterweeds currently harvested, it can be assumed that most of what is harvested is disposed of.

Table 2.13. Production from fish farms

Fish	Quantity produced (tons × 10^6)	Year	Area devoted to farms (ha)
Salmon and trout	13,400	1972	
Channel catfish	38,000	1975	18,809
Bait minnows		1975	16,290
Buffalofish	454		1,000

Source: Walsh, Don, 1978. *Aquaculture in the United States: Constraints and Opportunities,* Committee on Aquaculture, National Research Council, National Academy of Sciences, Washington, D.C.

2.4.3 Ecological Aspects

In many cases over the past century, degraded water quality has caused a decline of commercial fishing, especially in inland waters. In addition, production from many of the oceanic fisheries has peaked or is declining. Thus, the greatest potential for increasing the supply of fish protein may lie with production of freshwater fish, especially through aquaculture. Economical aquaculture and harvest from natural fisheries are dependent on adequate water quality.

As currently practiced, aquaculture itself can create significant sources of wastewater. Cultivating high concentrations of fish at rapid rates and using high-quality food create concentrated waste loads. Wastewater from ponds or raceways contains unused food, animal waste product, and chemicals and is often discharged without treatment. Many fish-farming operations are relatively small, so that the impacts of such discharge are frequently not serious. Nevertheless, the control of effluents from aquaculture requires continued effort by the industry and regulators (Walsh 1978).

2.5 Marine Systems

The nature of marine biomass harvesting and the complexity of the oceanic habitat render national boundaries without biological significance. Currents originating 1000 km away may be of predominant importance to a local fishery. A fleet may travel 10,000 km through international waters and fish along distant coasts before returning home to land its catch. Foreign ships routinely harvest U.S. waters, and many fish have oceanic migratory routes that allow a single stock to be under local exploitation by many nations (e.g., North Atlantic cod, North Pacific salmon, and tuna). The U.S. marine harvest must be considered in context within the world's fishery; therefore, a review of global marine biomass resources will precede the discussion of U.S. marine biomass and its utilization.

2.5.1 Global Marine Biomass

It has been estimated that marine phytoplankton fix a total of 1.9×10^{10} metric tons of carbon annually (Schaeffer 1965). This represents 1.9×10^{11} live-weight metric tons of primary productivity, assuming 10% of live weight is carbon. With a uniform trophic conversion efficiency of 15%, this primary productivity would support 2.8×10^{10} metric tons of herbivores, 4.2×10^{9} metric tons of first-stage carnivores, 6.4×10^{8} metric tons of second-stage carnivores, and 9.6×10^{7} metric tons of third-stage carnivores. Estimates of worldwide potentially harvestable marine biomass range from 100×10^{6} to 2000×10^{6} metric tons, depending on assumptions of trophic efficiency and harvest trophic level (Gullard 1971). World marine biomass harvest in 1974 totaled about 70×10^{6} metric tons (Bell et al. 1977), an increase from about 50×10^{6} metric tons in 1967.

The Food and Agricultural Organization divides the world's ocean resources into 15 regions. Over 60% of the world's harvested marine production can be accounted for by three regions: northwest Pacific, southeast Pacific, and northeast Atlantic. Most marine biomass (>85%) is harvested in the form of demersal (bottom living) and shoaling pelagic (inshore free-swimming) fish. Open-ocean areas provide only a small contribution to the total harvested fishery. In great contrast to land-based biomass systems, only about 1% of the world's harvested marine biomass is in the form of primary producers (seaweeds).

2.5.2 U.S. Marine Biomass

The marine biomass harvested by the United States ranks fifth internationally and accounts for about 4% of the world's harvest (Bell et al. 1977). Total U.S. marine landings in 1974 were 2.3×10^{6} metric tons and have remained relatively constant for the last 30 years. Landings of biomass for human consumption were 1.1×10^{6} metric tons in 1974, which represents the combined effort of about 160,000 fishermen working both onshore and in more than 15,000 vessels (Bell et al. 1977). The most productive U.S. regions are the Gulf of Mexico (with 40% of the U.S. catch by weight), the Pacific (27%), New England (12%), and the Chesapeake (11%). The catch in 1974 was divided between fish (80%) and shellfish (20%). Total landed value in 1974 was about $930 million. Table 2.14 describes the relationship between catch and distance offshore;

Tables 2.15 and 2.16 list the major harvested marine groups by weight and by value, respectively.

Table 2.14. Catch versus offshore distance (U.S. fishery)

Offshore distance (miles)	Total catch (%)
0-3	61
3-12	16
>12	12
International and foreign waters	11

Source: Bell, T. I., et al., 1977. *Fishery Statistics of the United States,* NOAA 78010407, National Marine Fisheries Service, National Oceanic and Atmospheric Admin., Rockville, Md.

Table 2.15. Harvested marine species by weight (U.S. fishery)

Species	Total by weight (%)
Menhaden	40.3
Tuna	7.9
Shrimp	7.5
Crab	6.7
Salmon	4.1
Anchovy	3.6
Flounder	3.3
Other	26.6

Source: Bell, T. I., et al, 1977. *Fishery Statistics of the United States,* NOAA 78010407, National Marine Fisheries Service, National Oceanic and Atmospheric Admin., Rockville, Md.

2.5.3 Seaweed Resources

The worldwide seaweed harvest for 1968 was about 0.9 × 10^6 metric ton (dry weight), although this estimate may be low due to a lack of data (Gullard 1971). Michanek (1975) estimates the harvest to be about 2 × 10^6 metric tons, with a worldwide natural harvest potential of 18 × 10^6 metric tons. Over one-half the reported seaweed harvest is consumed by Japan, where, in addition to its normal industrial uses, seaweed is a valuable staple of human consumption. Three genera comprise over 75% (by weight) of Japan's harvest: *Laminaria* (50%), *Undaria* (17%), and *Porphyra* (9%). Ten percent of the total Japanese marine harvest by value is in the form of seaweeds (5% by weight), but imports (mainly from Korea) are still necessary to meet domestic demand. A good portion of the seaweed harvest is derived from cultured stock, primarily *Porphyra, Laminaria, Undaria,* and *Monostroma* (on artificial substrates of net or rope) along the coastal region. Culturing one species of seaweed alone (*Porphyra* for "nori") is quite labor intensive, involving 60,000 Japanese families and using 9 million nets and floating blinds covering an area of over 200 km^2 (Michanek 1975). Seaweeds are cultured commercially to varying degrees, primarily in Japan, Korea, China, the Philippines, and Taiwan, with a pilot project to grow *Chondrus cripus* (for chemical feedstock) in Nova Scotia.

Table 2.16. Harvested marine species by value (U.S. fishery)

Species	Total by value (%)
Shrimp	19.4
Salmon	13.0
Tuna	12.8
Crab	9.1
Menhaden	7.2
American lobster	4.6
Other	33.9

Source: Bell, T. I., et al., 1977. *Fishery Statistics of the United States,* NOAA 78010407, National Marine Fisheries Service, National Oceanic and Atmospheric Admin., Rockville, Md.

The uses of kelp are quite varied. Potash, acetone, iodine, kelp meal (a food stock), and algin are some common products. Algin, in turn, is used in the manufacture of many items, including pharmaceuticals, cosmetics, candies, rubber, textiles, paints, and adhesives.

Chondrus cripus (Irish moss) is harvested from natural stands in Canada and Maine for use by the carrageenan industry (about 740,000 metric tons wet weight in 1969). *Ascophyllum* and *Laminaria* are also harvested for alginate production. During World War II, *Gracilaria* was harvested for agar production in North Carolina and Florida (Michanek 1975).

Currently the primary sources of seaweed in the United States are the kelp (brown alga) beds off the Pacific Coast. A comprehensive history of California kelp harvesting is found in Scofield (1959). The giant kelp (*Macrocystis pyrifera*) is the most important species. Bull kelp (*Nereocystis lutkeana*), found north of the Monterey Peninsula, has not been equally used. A small littoral marine red alga (*Porphyra perforata*) has been harvested for human consumption and shipped to China and to U.S. cities having oriental populations. The current harvest of *Porphyra* is estimated at 10 metric tons (Frey 1971). During World War II, when the agar supply from Japan was interrupted, seaweeds (Agarophytes) along the California coast were harvested and supplied eight to ten agar processing plants in the Los Angeles area. At present, there is no significant harvest of Agarophytes along the California coast (Frey 1971).

The rationale behind the choice of large brown kelps for intensive offshore biomass cultivation is given in Jackson and North (1973). They concluded that *Macrocystis* is the best choice for ocean farm cultivation in cool waters and that *Sargassum* is best for warmer waters. Work has been concentrated on developing an ocean farm concept viable for deployment off California which would use cool, deep, nutrient-rich upwelling water as fertilizer. The species of choice is therefore *Macrocystis*.

The three species of giant kelp are *Macrocystis pyrifera, M. angustifolia,* and *M. integrifolia* (Womersley 1954). Giant kelp occur over extensive regions of the Southern Hemisphere and along portions of the north Pacific coastline. The primary limiting conditions for establishment are availability of substrate (at a depth of about 2 to 40 m), cool temperatures (<25°C), and nutrient-rich water (high nitrate and phosphate concentration).

No true root system is present; nutrients are taken up directly from the surrounding medium, apparently to some extent over the entire surface of the plant. Attachment is by a holdfast to the substrate. Kelp fronds typically have a lifespan of from two to six months; losses are caused by temperature fluctuations, wave damage, grazing, senescence, etc. (North 1978). The upper portion of the canopy must continually be replaced by growth of new fronds. About 50 to 60% of the plant is contained in the top 3.5 m. Canopies can become very dense and block out more than 99% of ambient sunlight (Neushul 1971). Shaded portions of the plant are provided organic material transported down from the productive canopy (Parker 1971). Fronds near the base of the plant are specialized for spore production. A period of 2 to 2.5 years elapses from the time spores are released by an adult plant until the next generation plant is harvestable (North 1978). The giant kelp is a perennial with a life span of about 5 to 10 years (Frey 1971).

Productivity of natural kelp beds is estimated to range from 10 to 50 metric tons wet weight per hectare per year (Jackson and North 1973). California kelp beds are possibly nutrient limited for six to nine months per year, and yields from continuously fertilized and managed farms may greatly exceed natural rates (North 1978).

Beds of *Macrocystis pyrifera* cover about 200 km^2 (75 sq miles) off the California coast and comprise a standing crop of about 1.5×10^6 metric tons wet weight (North 1978). The total California harvest for 1975 was estimated at 143,000 metric tons wet weight (Pinkas 1977), with annual productivity estimated at 3×10^6 metric tons (Michanek 1975).

The total California harvest during the early years of this century was much greater, with the 1918 harvest estimated at about 360,000 metric tons (Frey 1971). By comparison, one proposed 40,470-ha (100,000-acre) kelp energy farm would cover an area of more than 388 km^2 (150 sq miles) and perhaps have an annual harvest of 3×10^6 dry metric tons or more, which is greater than the world's present total seaweed harvest.

References

Armson, K. A., 1977. *Forest Soils: Properties and Processes,* University of Toronto Press, Buffalo, N.Y.

Bell, T. I., et al., 1977. *Fishery Statistics of the United States—1974,* NOAA(NMFS)-78010407.

Bennett, G. W., 1970. *Management of Lakes and Ponds,* 2d ed., Van Nostrand Reinhold Co., New York.

Bethel, J. S., 1976*a*. *Reference Materials System: A Source for Renewable Materials Assessment,* NTIS PB 264 494, Committee on Renewable Resources for Industrial Materials, National Academy of Sciences, Washington, D.C.

Bethel, J. S., 1976*b*. *Renewable Resources for Industrial Materials,* NTIS PB 257 357, Committee on Renewable Resources for Industrial Materials, National Academy of Sciences, Washington, D.C.

Blankenhorn, P. R., T. W. Bowersox, and W. K. Murphy, 1978. "Recoverable Energy from the Forests: An Energy Balance Sheet," *Tappi* **61**(4): 57-60.

Brown, E. E., 1977. *World Fish Farming: Cultivation and Economics,* Avi Publishing Co., Westport, Conn.

Burwell, C. C., 1978. "Solar Biomass Energy: An Overview of U.S. Potential," *Science* **199:** 1041-48.

Carter, L. J., 1977. "Soil Erosion: The Problem Persists Despite Billions Spent on It," *Science* **198:** 409-11.

Council on Environmental Quality, Washington, D.C., 1978. *Environmental Quality: The Ninth Annual Report of the Council on Environmental Quality.*

Del Fosse, E. S., 1977. "Water Hyacinth Biomass Yield Potentials," pp. 73-100 in *Symposium on Clean Fuels from Biomass and Wastes, Jan. 25-28, 1977, Orlando, Fla.,* Institute of Gas Technology, Chicago, Ill.

Doering, O. C., III, 1977. "Future Competitive Demands on Land for Biomass Production," in *Symposium on Clean Fuels from Biomass and Wastes, Jan. 25-28, 1977, Orlando, Fla.,* Institute of Gas Technology, Chicago, Ill.

Douglass, J. E., and W. T. Swank, 1975. "Effects of Management Practices on Water Quality and Quantity: Coweeta Hydrologic Laboratory, North Carolina," pp. 1-13 in *Municipal Watershed Management—Symposium Proceedings,* U.S. For. Ser. Gen. Tech. Rep. NE-13, U.S. Department of Agriculture, Washington, D.C.

Fed. Regist., 1978. "Prime and Unique Farmlands," *Fed. Regist.* **43**(21): 4030.

Frere, M. H., et al., 1977. "Control of Nonpoint Water Pollution from Agriculture: Some Concepts," *J. Soil Water Conserv.* **32:** 260-64.

Frey, H. W., ed., 1971. *California's Living Marine Resources and Their Utilization,* California Dept. of Fish and Game, Sacramento, 148 pp.

Green, M. B., 1978. *Eating Oil—Energy Use in Food Production,* Westview Press, Boulder, Colo.

Gullard, J. A., ed., 1971. "The Fish Resource of the Ocean," *Fishing News,* UN/FAO, Ltd., Surrey, England.

Harte, J., and M. El-Gassier, 1978. "Energy and Water," *Science* **199:** 623-34.

Heichel, G. H., 1976. "Agricultural Production and Energy Resources," *Am. Sci.* **64:** 64-72.

Jackson, G. A., and W. J. North, 1973. *Concerning the Selection of Seaweeds Suitable for Mass Cultivation in a Number of Large, Open-Ocean, Solar Energy Facilities ("Marine Farms") in Order to Provide a Source of Organic Matter for Conversion to Food, Synthetic Fuels, and Electrical Energy,* Final Report by U.S. Naval Weapons Center to Energy Research and Development Administration.

Jahn, E. C., and S. B. Preston, 1976. "Timber: More Effective Utilization," *Science* **191:** 757-61.

Jensen, N. F., 1978. "Limits to Growth in World Food Production," *Science* **201:** 317-20.

Klass, D. L., 1976. "Wastes and Biomass as Energy Resources: An Overview of U.S. Potential," pp. 21-58 in *Symposium on Clean Fuels from Biomass, Sewage, Urban Refuse and Agricultural Wastes, Orlando, Fla., Jan. 27-30, 1976,* Institute of Gas Technology, Chicago, Ill.

Kochenderfer, J. N., and G. M. Aubertin, 1975. "Fernow Experimental Forest, West Virginia," in *Municipal Watershed Management Symposium Proceedings,* U.S. For. Ser. Gen. Tech. Rep. NE-13, U.S. Department of Agriculture, Washington, D.C.

Larson, W. E., R. F. Holt, and C. W. Carlson, 1978. *Residues for Soil Conservation, Crop Residue Management Systems,* American Society of Agronomy Spec. Publ. 31, Madison, Wis.

Lee, L. K., 1978. *A Perspective on Cropland Availability,* Agric. Econ. Rep. 406, U.S. Department of Agriculture, Washington, D.C.

Lockeretz, William, 1978. "The Lessons of the Dust Bowl," *Am. Sci.* **66:** 560–69.

Loehr, R. C., 1974. *Agricultural Waste Management—Problems, Processes, and Approaches,* Academic Press, New York.

McBride, E. J., 1978. Personal Communication, Conference on Environmental Aspects of Nonconventional Energy Resources, American Nuclear Society Topical Meeting, Denver, Colo., Sept. 26–29.

Megahan, W. F., 1972. "Logging, Erosion, Sedimentation—Are They Dirty Words?" *J. For.* **40**(7): 403–7.

Michanek, G., 1975. *Seaweed Resources of the Ocean,* Food and Agricultural Organization, Tech. Paper No. 138.

National Oceanic and Atmospheric Agency, 1977. *Fishery Statistics of the United States, 1974,* NTIS PB 277-796, National Marine Fisheries Service, U.S. Department of Commerce, Washington, D.C.

Neushul, M., 1971. "Submarine Illumination in *Macrocystis* Beds," chap. 10 in *Biology of Giant Kelp Beds* (Macrocystis) *in California,* W. J. North, ed., Cramer, Lehre, Germany.

North, W. J., 1978. "A General Description of *Macrocystis,*" in *Handbook of Solar Biosources,* vol. I, *Basic Principles,* C. C. Black and A. Mitsui, eds., CRC Press, Inc., Cleveland, Ohio.

———, in press. "Evaluation, Management, and Cultivation of *Macrocystis* Kelp Forests," *Symp. on Chilean Algae, Santiago, Chile, Nov. 21–24, 1978.*

Parker, B. C., 1971. "Studies of Translocation in *Macrocystis,*" in *Biology of Giant Kelp Beds* (Macrocystis) *in California,* W. J. North, ed., Cramer, Lehre, Germany.

Pimentel, D., et al., 1973. "Food Production and the Energy Crisis," *Science* **182:** 443–49.

———, 1975. "Energy and Land Constraints in Food Protein Production." *Science* **190:** 754–61.

Pinkas, L., 1977. *California Marine Fish Landings for 1975,* California Dept. Fish and Game Bull. 168, Sacramento, Calif.

Schaeffer, M. B., 1965. "The Potential Harvest of the Sea," *Trans. Am. Fish. Soc.* **94**(2): 123–28.

Schiff, S. D., 1979. "Land and Food: Dilemmas in Protecting the Resource Base," *J. Soil Water Conserv.* **34:** 54–59.

Scofield, W. L., 1959. "History of Kelp Harvesting in California," *Calif. Fish Game* **45**(3): 135–57.

(SCS) Soil Conservation Service, 1966. *Land Capability Classification,* Agricultural Handbook 210, U.S. Department of Agriculture, Washington, D.C.

———, 1977. *USDA Potential Cropland Study,* SCS Statistical Bulletin 578, U.S. Dept. of Agriculture, Washington, D.C.

Stanford Research Institute, Menlo Park, Calif., 1976. *Crop, Forestry, and Manure Residue Inventory,* vols. 1–8.

Steinhart, J. S., and C. E. Steinhart, 1974. "Energy Use in the U.S. Food System," *Science* **184:** 307–16.

Stephens, G. R., and G. H. Heichel, 1975. "Agricultural and Forest Products as a Source of Cellulose," *Biotechnol. Bioeng. Symp.* **5:** 27–42.

Tebo, L. B., Jr., 1955. "Effects of Siltation, Resulting from Improper Logging, on the Bottom Fauna of a Small Trout Stream in the Southern Appalachians," *Prog. Fish Cult.* **17:** 64–70.

Thomas, G. W., and J. W. Gilliam, 1977. "Agro-Ecosystems in the USA, Cycling of Mineral Nutrients in Agricultural Ecosystems," *Agro-Ecosystems* **4:** 182–243.

Tillman, D. A., 1978. *Wood as an Energy Resource,* Academic Press, New York.

Train, R. E., 1978. "The Environment Today," *Science* **201:** 320–24.

(USDA) U.S. Department of Agriculture, Washington, D.C., 1967. *Geographic Forest Types Used in the Forest Survey,* For. Serv. Handbook 4813.1, sect. 74.

———, 1973. *The Outlook for Timber in the U.S.—1970,* For. Serv. Res. Rep. 20.

———, 1974. *Our Land and Water Resources—Current and Prospective Supplies and Uses,* Economic Research Service, Misc. Publ. 1290.

———, 1977. *The Nation's Renewable Resources—An Assessment, 1975,* For. Serv. Res. Rep. 21.

———, 1978. *Agricultural Statistics.*

(USEPA) U.S. Environmental Protection Agency, 1978. *Preliminary Environmental Assessment of Biomass Conversion to Synthetic Fuels,* EPA-600/7-78-20F, Washington, D.C.

(USFS) U.S. Forest Service, 1975. *Municipal Watershed Management—Symposium Proceedings (September 1973),* For. Serv. Gen. Tech. Rep. NE-13, U.S. Department of Agriculture, Washington, D.C.

Walsh, D., 1978. *Aquaculture in the United States: Constraints and Opportunities,* National Academy of Sciences, Washington, D.C.

Water Resources Council, Washington, D.C., 1974. *Water Requirements, Availabilities, Constraints, and Recommended Federal Actions,* Fed. Energy Admin. Proj. Independence Blueprint, Final Task Force Report.

Wolverton, B. C., and R. C. McDonald, 1979. "The Water Hyacinth: From Prolific Pest to Potential Provider," *Ambio* **8**(1): 2–9.

Womersley, H. B. S., 1954. "The Species of *Macrocystis* with Reference to Those on South Australian Coasts," *Univ. Calif. Berkeley Publ. Bot.* **27**(109).

3

Proposed systems for production and harvest of biomass for energy

The biomass sources from which energy could be harvested comprise three distinct categories of material—wastes, residues, and new production. Wastes are biomass materials that must currently be disposed of at an economic cost. Residues are products that have a current use and, therefore, some value, although use may not entail active collection, transportation, or sale. New production denotes energy crops, that is, biomass specifically grown for its energy content. This section considers briefly the proposed production and harvesting systems for materials from each of the major production fields.

3.1 Forestry and Silviculture

3.1.1 Mill Wastes

Wood processing operations include sawmilling for production of lumber, processing for production of plywood and composition products, and wood pulping for production of paper and fiber products. Each operation produces various types of primary mill wastes; wood pulping also produces spent pulping liquor, a secondary waste.

Primary wood waste consists of bark, chips, sawdust, planer shavings, sander dust, and other material. Bark is generated in the debarking operation at many plants, although debarking is not universal. Some sawmills simply produce slabs that include bark, and some pulp mills can accept a limited proportion of whole-tree chips, including bark. Sawdust from woodcutting is produced in all sawmills; planer shavings and sander dust are produced in the finishing of lumber. Plywood production produces coarse residues such as core logs and trim pieces as well as some sawdust. Coarse residue, large debris, defective products, and unusable wood are "hogged" (chipped or ground up) to provide chips for wood pulping, or "hogged fuel." In the production of lumber and plywood, 55 to 66% of the total weight of logs brought to the mill appears as refuse, including bark, slabs, sawdust, and chips (Howlett and Gamache 1977*a*).

Spent pulping liquor, the "black liquor" that results from dissolving wood chips during pulp making, contains the energy-rich component of wood (lignin) and valuable pulping chemicals such as sodium. Spent pulping liquor is concentrated to about 65% solids and is burned in energy-recovery boilers from which pulping chemicals can be recovered. This process supplied 76% of the wood-derived energy and 37% of the energy consumed by the pulp and paper industry in 1976 (Tillman 1978, citing Duke and Fudali 1976). Wood pulping produces mostly bark as a waste and uses a greater proportion of logs than lumber or plywood production. Much of this bark is presently being used by the paper industry in boilers.

The Forest Service (USDA 1973) reported that, in 1970, 75% of the wastes at primary wood processing plants were used. About 56%

of the wood wastes generated were used for nonenergy products, primarily for pulp, whereas 19% were either used onsite for fuel or sold as fuel. The remaining 25% of residues were unused. Data on the regional distribution of mill wastes are given in Table 3.1. The greatest quantities are produced in the Pacific Northwest, the southcentral region, and the Southeast. The proportion of total waste used in 1970 varied from 85% (Pacific Northwest) to 44% (southern Rocky Mountains).

Table 3.1. Wood and bark wastes from processing, by region in 1970

Region[a]	Total wastes generated (DTE $\times 10^6$)[b]	Proportion used (%)
Northeast	6.6	65
North central	6.4	67
Southeast	11.4	61
South central	16.7	72
Pacific northwest	27.8	85
Pacific southwest	8.8	63
Northern Rocky Mountains	6.6	68
Southern Rocky Mountains	1.8	44
	$\Sigma = 86.1$	$\bar{x} = 72$

[a]Regions contain the following states: northeast — Conn., Maine, Mass., N.H., R.I., Vt., Del., Md., N.J., N.Y., Pa., and W.Va.; north central — Mich., Minn., N.D., S.D., Wis., Ill., Ind., Iowa, Kan., Ky., Mont., Neb., and Ohio; southeast — N.C., S.C., Va., Fla., and Ga.; south central — Ala., Miss., Tenn., Ark., La., Okla., and Tex.; Pacific northwest — Ore., Wash., and Alaska; Pacific southwest — Calif. and Hawaii; northern Rocky Mountains — Idaho, Mont., western S.D., and Wyo.; southern Rocky Mountains — Ariz., Colo., Nev., N.M., and Utah.

[b]DTE = dry ton (English) equivalent (1 DTE = 0.91 dry metric ton).

Source: Howlett, K., and A. Gamache, 1977. Table 2-17, p. 33, in *Silvicultural Biomass Farms*, vol. 6, *Forest and Mill Residues and Potential Sources of Biomass*, Tech. Rep. 7347, Mitre Crop., McLean, Va.

Although no accurate data on the current utilization of mill wastes are available, it seems likely that the percentage of wastes currently unused is lower than the 26% reported in 1970. Steinbeck (1978) reports that the use of wood for energy is accelerating in the southeastern United States at such a pace that in some areas "old sawdust piles have disappeared and competition for sawmill residues is becoming stiffer." In all the timber-producing areas of North America, the use of mill waste for energy has recently increased. Examples include use of bark in pulp mill boilers (Calhoun, Tennessee, and Rumford, Maine), steam production for a textile mill using woodchips purchased from 30 small sawmills (Alexander City, Alabama), cogenerating electricity and space heat from hogged fuel from nearby mills (Eugene, Oregon), and plywood veneer drying by use of trim scraps (Omak, Washington) (Biomass Energy Institute, Inc. 1978).

The schemes proposed for future use of mill waste are simply an extension of the past rapid trend. However, large volumes of wood waste which remain unused are those that are (1) remote from a boiler, (2) uneconomical to handle or transport (for instance, caused by low density or high moisture content), or (3) of poor quality (e.g., having low fuel value or high impurity content). Schemes to use mill wastes may involve installation of woodburning boilers or modifications to existing boilers to accept mill wastes. An alternative for small mills or sites where boilers are not needed is to transport mill wastes to some suitable boiler. Mill wastes can be handled with front-end loaders, scrapers, or similar equipment and can be moved from disposal site to point of use by trucks, conveyer belts, or air-blown transport. Because many larger mills are already using mill wastes, the increased use of these materials will undoubtedly involve some of the smaller mills. Operators at these sites may be able to sell sawdust to nearby larger mills, and the market for slabs as firewood for homeowners will probably increase.

As a specific example of proposals to increase use of wood waste, the Weyerhauser Corporation is developing improved methods of collecting, cleaning, and burning debris in the log yards. Boilers are being modified to accept a wider range of wood-particle sizes and qualities; for example, cyclone burners are being added to burn sander dust. However, Weyerhauser perceives the wood-energy resource base as limited in the long term; therefore, energy systems that can use alternative fuels (such as coal) are being planned (Jamison 1977).

The Georgia-Pacific Corporation examined prospects for procuring mill wastes for proposed or existing boilers in northern New England and southern Arkansas (Howlett and

Gamache 1977*a*). Many of the mills surrounding Gilman, Vermont, reported that waste piles existed, that wastes were available, and that prospects were good for continued availability of mill wastes. This situation resulted from (1) increasing production from the region's forests and (2) declining use of sawdust and shavings as bedding for dairy cattle. However, the prospects for getting additional mill wastes for Georgia-Pacific's boilers in Crossett, Arkansas, were less promising. Mill wastes were found to be in demand as boiler fuel at other pulp mills and for use in particle-board mills.

For the next decade the availability of waste wood will be changing and uncertain. The resource will be limited by the industry trend toward using a greater fraction of the tree for composite materials and pulping rather than for energy. Additionally, lumber mills are using improved technologies, such as thinner blades, which will reduce production of sawdust and planer shaving wastes (Benemann 1978). Competition for mill wastes will intensify, and some nonforest industries may penetrate the market if they are located close to the waste materials [North Central Forest Experiment Station (NCFES) 1978]. As complete utilization of waste materials is approached, demand for forest residues and biomass from marginal forests will increase.

3.1.2 Forest Residues

Forest residue includes material from forest management and logging residues. Some forest management activities that produce residues are intermediate cuttings, understory removal, and cull tree removal. These practices are intended to remove plant growth that competes with the desirable timber species. Thinnings may occur in stands not past the sapling stage, but the amount, size, and dispersal of the residues generated are not conducive to effective collection, transportation, and merchandising using conventional methods (Howlett and Gamache 1977*a*). Thinning of older stands, however, may generate merchantable timber or residues suitable for energy uses, and income from sale of residues can enhance a forest manager's opportunity to improve a timber stand. Two methods that might improve feasibility of using residues from thinning for energy are (1) row thinning, in which all stems are removed in a 2- to 3-m swath leaving narrow bands of trees, and (2) understory removal by specialized machinery to replace controlled burning or herbicide application (Howlett and Gamache 1977*b*). However, the lack of equipment capable of collecting such forest residues economically is a major barrier to their increased use at present.

Logging residues present a more promising prospect than thinnings for fuel supply, and commercial logging operations are currently the most important single source of forest residues. Because most forest product manufacturing requires only the bole or stem of the tree, considerable volumes of tops, branches, and stump-root systems remain after harvest. Brush, foliage, and rough, rotten, and damaged timber all contribute to logging residues.

The merchantable stem usually constitutes about 75 to 90% of the aboveground tree biomass, depending on the tree size, species, and growing conditions (Howlett and Gamache 1977*a*). However, because of rough trees, damaged timber, etc., residue mass varies from about 25 to 45% of wood cut. Table 3.2 indicates the quantities of residue generated per unit area for regions of the United States. The rate of residue production per hectare is highest in the Pacific Coast states and is lowest in the lake, central, and gulf states.

Obviously, not all the residues left in the forest after logging are available from an economic standpoint. Logging residues are scattered over a wide area when they are produced, and some fraction of logging slash is unrecoverable under almost any circumstance. In some cases residues are already collected. The reasons for residue collection include the need to (1) clear land for replanting, (2) remove diseased timber, (3) reduce fire hazard by removing excess wood, and (4) improve the appearance of cleared land following harvest. The collection of forest residues is sometimes called yarding unmerchantable material and often entails considerable expense. The wood is sometimes burned onsite, which may contribute to air pollution problems. Increasingly, however, the wood is chipped or hogged and used as fuel or pulp feedstock. In national forests the Forest Service sets standards for slash disposal

Table 3.2. Estimates of logging residues by region

Region	Average harvest volume[a] [m^3/ha (ft^3/acre)]	Average residues per area harvested [dry metric ton equivalent/ha (DTE/acre)][b]	
		Aboveground	Stump-root systems
New England	75.2 (1076)	18.6 (8.3)	21.9 (9.8)
Middle Atlantic	66.4 (949)	24.7 (11.0)	23.1 (10.3)
Lake states	54.3 (776)	12.1 (5.4)	15.9 (7.1)
Central states	42.5 (607)	13.9 (6.2)	14.1 (6.3)
South Atlantic	64.1 (916)	23.1 (10.3)	21.3 (9.5)
East Gulf	49.3 (705)	12.6 (5.6)	14.4 (6.4)
Central Gulf	54.5 (779)	14.4 (6.4)	16.4 (7.3)
West Gulf	57.1 (817)	15.0 (6.7)	17.1 (7.6)
Pacific northwest	237.4 (3395)	53.8 (24.0)	62.4 (27.8)
Pacific southwest	188.7 (2698)	38.8 (17.3)	49.6 (22.1)
Northern Rocky Mountains	119.5 (1709)	21.3 (9.5)	30.3 (13.5)
Southern Rocky Mountains	83.4 (1193)	15.3 (6.8)	21.8 (9.7)
Total	86.5 (1236)	20.4 (9.1)	24.4 (10.9)

[a]Estimates based on average timber inventories per acre (growing stock plus nongrowing stock) in 1970 and average proportions of inventories harvested in 1970.

[b]DTE = dry ton equivalent.

Source: Howlett, K., and A. Gamache, 1977. Table 3-14, p. 77, in *Silvicultural Biomass Farm*, vol. 6, *Forest and Mill Residues as Potential Sources of Biomass*, Tech. Rep. 7347, Mitre Corp., McLean, Va.

which vary for different parts of the country and change over time (Zerbe 1976).

The economics and equipment for collecting forest residues are critical factors. In the old growth stands of the Pacific northwest, large quantities of large-size residues can often be collected, perhaps economically, with conventional equipment such as skidders and loaders. Most forest residues, however, are smaller in size and awkward to handle, which makes economical collection difficult.

Some existing equipment is adaptable to residue collection, but the use of skidders, larger drum cable systems, stump splitters, and semimobile or mobile chippers entails high costs. Equipment that could be adapted to residue processing (Table 3.3) includes chippers without collecting capabilities (e.g., Chipharvester, Precision tree harvester, Ecolo chipper) and mobile hogs that leave chips, splinters, or shredded wood on the forest floor in windrows (e.g., Tree eater, Hydro-Ax, and Klear Way) (Howlett and Gamache 1977*b*). Koch and McKenzie (1976) proposed the development of a mobile hog that is specifically designed for collection and reduction of forest residues. Steep or rough terrain, however, will impose limitations on large pieces of equipment.

3.1.3 Marginal Forests

Another potential source of wood for energy is biomass from standing forests that would not ordinarily be cut. The United States has large areas of low-grade forests that receive little or no management because of poor site quality and fragmented ownership. Wood from such forests constitutes a special category of residue. Although it is not currently economically feasible to log these marginal forests for timber or pulpwood, they could be cut to supply fuel wood and/or to remove low-quality forests and replace them with higher-quality forests. (The term "marginal forests" is unfortunate because the lands are marginal only from an economic standpoint. Marginal forests may have significant value for recreation, wildlife, erosion control, or other uses.) Marginal forests could make a large contribution to near-term harvest of wood for energy, especially in the eastern United States, where low-grade

Table 3.3. Equipment adaptable to collecting and/or reducing forest residues

Equipment	Slope limitation[a] (%)	Size limitation Diameter (in.)	Length (ft)	Cost[b] ($/DTE)	Support equipment needed[c]	Manufacturer
Morbark Chipharvestor	Limit of skidder	22	None	6.60[d]	Skidders and chain saws	Morbark Ind., Inc., Winn, Mich.
Precision tree harvester	Limit of skidder	22	None	6.60[d]	Skidders and chain saws	Precision Chipper Corp., Birmingham, Ala.
Nicholson ecolo chipper	Limit of skidder	24	None		Skidders and chain saws	Nicholson Mfg. Co., Seattle, Wash.
Tree eater	20	10	None	5.50 at 11 DTE/h	None	Tree Eater Corp., Gurdon, Ark.
Wagner-Bartlett stump splitter-remover	Limit of loader	96	None	3.50/stump	Mounted on loader	Wagner Mfg. Co., Portland, Ore.
National Hydro-Ax	30	6	None	4.40 at 11 DTE/h	None	National Hydro-Ax, Inc., Owatoma, Minn.
Kershaw Klear Way	25	6	None	4.40 at 11 DTE/h	None	Kershaw Mfg. Co., Inc., Montgomery, Ala.

[a]Based on working performance on firm soils.

[b]Includes all known, fixed and maintenance, move in and out, and necessary personnel; DTE = dry ton (English) equivalent (1 DTE = 0.91 dry metric ton).

[c]*Does not* include equipment needed for accumulation of reduced residues.

[d]Chipping cost only. Total cost estimated at $20.00 per dry metric ton.

Source: Howlett, K., and A. Gamache, 1977. Table 3-18, p. 82, in *Silvicultural Biomass Farms*, vol. 6, *Forest and Mill Residues as Potential Sources of Biomass*, Tech. Rep. No. 7347, Mitre Corporation, McLean, Va. Based on data from Harrison, R. T., 1975. *Slash-Equipment and Methods for Treatment and Utilization*, Equipment Development and Test Report 7120-7, U.S.D.A. Forest Service, Equipment Development Center, San Dimas, Calif.

forest lands are abundant. Equipment for this type of harvest is available or under development and includes feller-bunchers, skidders, chippers, mobile hogs, and cable logging systems.

Several specific proposals for harvesting unused forest biomass for energy production have recently appeared. A 1978 study by NCFES in St. Paul, Minnesota, compared the energy available from forest residues with the demands of the Mead Corporation pulp mill in Escanaba, Michigan, on the upper peninsula. Based on the data collected in the study (NCFES 1978), a hypothetical scheme for supplying wood energy to the Escanaba mill might involve harvesting as much as 1.9 million tons of saplings, tops, branches, rough trees, and rotten wood at an average cost of $13.60 per ton in 1980. This quantity of forest residue would render the mill completely energy self-sufficient but would require modification or replacement of several boilers and other equipment. Depending on the forest and its condition, harvest would involve (1) clear-cutting, (2) thinning, and (3) selection cutting for sawlogs, followed by relogging to recover residues.

Harvesting operations are depicted in Fig. 3.1. Feller-bunchers shear the trees near the stump and assemble them for grapple-skidders, which carry or drag bunches to chippers. Bunches are disassembled at the chippers and sorted to remove merchantable sawlogs, and residue material is fed to the chipper. Both rubber-tired and tracked vehicles are available for these jobs. Chips would be trucked to the mill and perhaps dried and fired to provide process heat. Such increased utilization has already been under way for fiber recovery.

Another proposal for harvesting unused forests involves a wood-fired power plant (WFPP) in northern New England, where facilities have been proposed for Burlington, Vermont, and southeastern Maine. The state of Vermont commissioned a study of wood availability in northern Vermont (Hewett 1978). The region's forests resulted largely from abandonment of farmland and pasture and are predominantly unmanaged. They have also been subjected to poor harvest practices such as "high-grading" or selective removal of high-quality trees. As a result, forests in the nine-county study region have

Fig. 3.1. A typical whole-tree field chipping system. Source: North-Central Forest Experiment Station, 1978. *Forest Residues Energy Program. Final Report*, TID-28416, U.S. Department of Energy, Washington, D.C., Fig. 2.4, p. 28

declined in quality and show relatively poor growth rates and large proportions of non-merchantable timber.

Around the proposed Burlington WFPP, 8% of the land is owned by federal and state governments, about 60% by private individuals, and about 30% by corporations. Forty-five percent of the area was judged unavailable for harvest because of steep slopes (>15%) and distance from roads. Hewett (1978) suggests that private owners might open one-third of their land to whole-tree harvest and that the region could produce roughly 168% of the wood required for a WFPP. Harvest would involve whole-tree chipping, adapted to the steeper slopes of northern Vermont, perhaps with the use of smaller, more maneuverable equipment. Merchantable logs would be removed before chipping, adding to the profitability of the harvest. Transport to the power plant would be by truck.

3.1.4 New Production

Because wastes and residues constitute a somewhat limited resource base for energy production, the concept of silvicultural energy farms (SEFs) has been advanced. At an SEF, trees would be intensively managed to produce an energy crop. The SEF concept has been studied extensively by the Mitre Corporation. As envisioned in the Mitre study (Howlett and Gamache 1977*b*), an SEF would use tree species (1) selected for high yield, (2) well suited to the local site conditions, and (3) responsive to management. Species that would coppice, or sprout from stumps, would be desirable to eliminate the need to reestablish the stand after harvest. Short-rotation management (four to eight years) would probably be used to get maximum yield, rapid return on investment, and increased efficiency through mechanization. Establishing an SEF would involve intensive site preparation and planting of high-density stands every few rotations. Management would be intensive and would include fertilization, weed control, pest control, and perhaps irrigation. Harvesting might involve a self-propelled harvester that would continuously shear a swath of small trees, chip them, and blow the chips into a wagon.

An SEF would represent a substantial departure from current forestry practices. Although tree plantations exist, particularly in the southern United States, trees are widely spaced to allow growth to the large sizes suitable for production of solid wood products or fiber. Rotation periods are from 20 to 70 years. Industries currently procure wood from public lands, industry-owned lands, and private lands, and the mix of sources varies greatly. The intensive management required for an SEF would point to the desirability of large, relatively contiguous tracts under uniform ownership or cooperative management. The size of an SEF would probably result from a compromise between the desirability of a large tract and the difficulty of procuring suitable sites and/or cooperators.

Numerous forest tree species are being evaluated as candidates for biomass production, including American sycamore, *Eucalyptus,* loblolly pine, red and black alder, sweetgum, cottonwood, black locust, sugar maple, and others. The criteria being evaluated include (1) rapid juvenile growth, (2) ease of establishment and regeneration, (3) freedom from major insect and fungal pests, and (4) tolerance of variable or poor site conditions. Another important factor is the susceptibility of species to further improvement by selection and genetic manipulation.

Extensive site preparation is considered to be the key to the success of any short-rotation forestry scheme. Slash disposal, soil scarification by disking, leveling and draining, and control of unwanted new growth with herbicides may be employed. Seedlings or cuttings might be planted by machine in densities from 1,200 to 10,000 stems per acre (compared with 200 to 700 stems per acre for conventional forest plantations). Until complete canopy closure occurs, control of competing plants would involve mechanical cultivation or chemical control. Fertilization would be required not only to enhance growth, but also to maintain fertility through multiple rotations. Protection from fire, insects, and disease would require use of firebreaks, fire control, and biocides; irrigation systems might be necessary in drier climates to ensure maximum growth. A silage-type harvester could collect the biomass after very short rotations. Harvested chips would be transported by truck to the point of use.

The Mitre study (Inman 1977) developed a conceptual production design for an SEF and projected development at ten U.S. sites. The SEFs were sized to produce 250,000 dry ton equivalents (DTE) (~226, 790 day metric tons) per year, approximately enough fuel for a 50-MW power plant. The current productivities were estimated as 11 to 29 metric tons/ha-year (5 to 13 DTE/acre-year), rising in the future to 22 to 49 metric tons/ha-year (10 to 22 DTE/acre-year) with improved productivity and technology. The rotation period was projected as 6 years with a 30-year lifetime per planting. Land requirements for one SEF ranged from 8,500 to 28,700 ha (21,000 to 71,000 acres) with current productivity, or 5,260 to 14,570 ha (13,000 to 36,000 acres) with future productivity.

3.2 Agricultural Crops

3.2.1 Agricultural Residues

Crop residues include stems, leaves, roots, chaff, and any other plant parts that remain after agricultural crops are harvested. Traditionally, these materials have functioned in maintenance and protection of soil. An extremely minor fraction (2%) of residue is considered to be waste (Sect. 2.2.4).

The great attraction of crop residues for fuel recovery is that the cost of producing the residue is included in the cost of producing the primary crop, and the only additional expenses are for harvesting, storage, and transportation of the residue itself. These expenses may, however, be considerable, given the bulky form of residue materials and the large areas over which they are initially distributed. The energy density of field residues is also low, only 10 to 20% of that of wood (Burwell 1978). Nevertheless, the overall energy content of residues is large; for example, the annual U.S. production of cereal straw residues is estimated to contain a total of 2.2 EJ (2.1 quads) and the energy content of cornstalks is estimated at 1.9 EJ (1.8 quads) (Burwell 1978).

In contrast to forestry, the development of equipment specifically for collecting and handling residues from agriculture has received little attention (Stanford Research

Institute 1977). Efforts in agriculture have been traditionally directed toward *minimizing* the amounts of residue harvested with the primary crop. A new orientation to the deliberate maximization of residue collection will require basic changes in equipment and procedure as well as economic incentives. Experimentation to produce equipment with greater collection efficiency has been inhibited by the reluctance of farmers to invest capital in high-risk, marginal revenue-return efforts. This problem could be circumvented, in part, by reliance on custom farm services to harvest residues (Tinberg and McElroy 1978). Variations in yield and climatic conditions at harvest time could also have adverse effects on collection success. For example, corn stover must be harvested following the corn grain, which would necessarily be late in the year during a period in which inclement weather is likely to inhibit equipment use and escalate costs.

Certain procedures necessary for harvesting, regardless of the type of crop, include mowing, raking, baling, and moving from field to storage, especially for seasonal commodity residues. Loading and transportation to the point of use must also be included as cost and energy considerations.

Agricultural breeding programs have also traditionally been oriented toward minimizing plant parts not usable as food. Plant genetics will need to be applied to reverse this selection process if maximization of residue production becomes desirable.

Crop residues can be divided into four categories, based largely on moisture content and handling characteristics (Stanford Research Institute 1977); these categories are (1) high-moisture field residues (moisture content 40 to 80%) such as corn and sorghum stovers; (2) low-moisture residues (moisture <40%) such as straws; (3) vegetable field residues (moisture >80%) such as tomato and potato vines and sugar beet tops; and (4) mill residues from crop processing facilities (e.g., bagasse, sugar beet pulp, cotton gin trash, and rice and sunflower hulls).

Generally, as moisture content of residues increases, conversion efficiency decreases. Residues with a moisture content greater than 50%, for example, would have to be predried or fired with supplementary fuels to be suitable for combustion. Feedstocks with a moisture content below 50%, however, could be fed directly for production of synthetic natural gas. Anaerobic digestion requires low-solids feedstocks (80 to 97% water). Different residues are most appropriate for different conversion processes (Chap. 4). Efficiencies and cost factors vary, however, for different feedstocks, processes, and end products. This may have the effect of channeling greater development effort into a selected few of the possible fuel-production schemes.

Local or onsite use of residues is considered a sound concept (Miles 1977). Where feasible or desirable, this concept tends to simplify logistics problems by minimizing transportation and storage costs as well as by conserving energy. The success of onsite use is already evident in the sugarcane, pulp and paper, and forest products industries.

3.2.2 Energy Crops

Nonforestry plant species that may be cultivated for their energy content fall into three basic categories: (1) the carbohydrate crops, including sugarcane, sweet sorghum, corn, and other common sources of fermentable sugars; (2) grasses and grains, which could be the source of materials for combustion or synthetic gas production; and (3) hydrocarbon crops, such as guayule, jojoba, *Euphorbia,* and others that produce oils, waxes, latexes, or other special substances. Some plant species provide several plant parts, which makes them suitable for multiple purposes; for example, corn can be a source of both fermentation (grain) and combustion/digestion (stover) feedstocks. The major factor common to all these species is that their production, whether on a large or a small scale, will entail application of agricultural practices.

3.2.2.1 Carbohydrate crops. Plants that produce fermentable sugars and starches are potential sources of ethanol for motor fuel. The use of sugarcane, sugar beets, and sweet sorghum, which provide sugars directly, would be most efficient. The starch crops require some initial treatment to produce fermentable sugars.

Sugarcane is a tropical grass with narrow physiological tolerances to temperature and water availability. The geographical area of the United States suitable for its cultivation is therefore extremely limited and confined largely to Florida, Louisiana, Texas, and Hawaii. The knowledge and technology for successful sugarcane production are, however, available and well understood. Sweet sorghum has the potential to grow anywhere that cotton, corn, grain sorghum, sugar beets, and sugarcane grow (i.e., over a huge area). At present, however, the area devoted to this crop is negligible (Battelle Columbus Laboratories 1976), and no satisfactory harvesting system is currently available (Battelle Columbus Laboratories 1978). Hence, large-scale cultivation of sweet sorghum is questionable at present because of technological, economic, and sociological uncertainties regarding its acceptance as a common cash crop. Sugar beet cultivation is widespread, but the use of this plant for chemical feedstocks has been judged uneconomical (Battelle Columbus Laboratories 1977*a*); sugar beet materials are considered more valuable as cattle feed.

Sugarcane and sweet sorghum are currently grown and bred exclusively for human consumption. Production for chemical feedstocks will require the addition of a considerable land area. A Battelle Columbus Laboratories study estimates that 59 million metric tons (65 million tons) of fermentable sugars (sufficient for 8.5 billion gallons of ethanol for a 10% "gasohol" blend for 75% of U.S. motor fuel) could be produced by 1998 if the land area of cane and sorghum harvested were increased from 0.32 million ha (0.78 million acres) (1978 level) to 6 million ha (14.5 million acres) (Battelle Columbus Laboratories 1977*a*). The details of the suggested distribution of this increase are given in Table 3.4. The land in Florida thought to be convertible to sugarcane is now in palmetto and pine. In Louisiana, sugarcane production would have to be increased by either shifting acreage from cotton and soybeans or by draining coastal wetlands. The Rio Grande Valley of Texas would be

Table 3.4. Volume goals from sugar crops for year 1998

Crop/area	Harvested (acres × 10^6)		Biomass (tons/acre/year)		Total biomass (tons × 10^6)	Biomass available for energy[a] Tons × 10^6	Biomass available for energy[a] Quads	Fermentable sugars (tons × 10^6)[b]
	1998	1978	1998	1978			1998	
Sugarcane								
Florida, west of Belle Glade	1.00	0.31	20[c]	10.9	20.0	16.6	0.25	5.6[d]
Louisiana	0.70	0.33	16	7.5	11.2	8.7	0.13	2.6
Texas	0.4	0.03	20[c]	11.7	8.0	7.7	0.12	2.4
Mississippi/Alabama	0.10		12		1.2	1.2	0.02	0.3
Hawaii	0.22[e]	0.11	25	15.9	5.5	2.0	0.03	0.6
Subtotal	2.4	0.78	19	12.9	46.0	36.0	0.60	11.6 = 1.5 billion gal EtOH
Sweet sorghum								
Florida	1.00		20[c]		20.0	20.0	0.30	5.4[f]
Louisiana	0.50		15		7.5	7.5	0.11	2.6
Texas	0.30		18[c]		5.4	5.4	0.08	1.5
Mississippi/Alabama	0.30		15		2.3	2.3	0.03	0.8
Midwestern states	10.00		15		150.0	150.0	2.25	40.4
Subtotal	12.1				185.2	185.2	2.8	51 = 7 billion gal EtOH
Total	14.5				231	221	3.4	63 = 8.5 billion gal EtOH

[a]After deducting 1976 sugarcane for food use.

[b]One ton of fermentable sugars is capable of producing about 140 gal of ethanol or 110 gal of 2,3-butanediol.

[c]Optimized yields on sandy soil.

[d]Estimated percentage fermentable sugars as percentage of total dry weight millable sugarcane: Florida, 45%; Louisiana, 40%; Texas, 40%; Mississippi/Alabama, 40%; and Hawaii, 49%.

[e]Increased acreage represents more frequent harvest rather than new land development.

[f]Estimated percentage fermentable sugars as percentage of dry weight millable sweet sorghum: Florida, Texas, and Midwest states, 35%; Louisiana, Mississippi, and Alabama, 45%.

Source: Battelle Columbus Laboratories, 1978. *Potential Benefits and Technology Development Costs for Sugar Crops as an Energy Resource*, topical report prepared for U.S. Department of Energy, FFB Branch.

suitable if nonsaline well or river water were available for irrigation. Hawaiian production would have to be increased almost purely by improvements in cultural practices because little additional land is available for expansion. If cane and sorghum economics were favorable, a projected revival of these crops could also potentially occur in Mississippi and Alabama. This revival would require the displacement of cotton and soybeans. Production of sweet sorghum, as projected for 1998, entails acceptance as a large-scale crop and, at least for the Midwest, inclusion into the existing corn/soybean crop rotations.

Agricultural technology developments necessary for successful production of fuel ethanol will involve three major categories of endeavor: plant variety development, cultural practices, and agricultural engineering. Virtually all ground rules for the selection of optimal varieties are substantially different when biomass fuels, rather than edible foodstuffs, are the goal. Cold tolerance, maturation characteristics, and increased fiber production must replace strict maximization of sucrose content (Alexander 1977). The sought-for qualities will probably require extensive selection programs outside of the conventional plant varieties. Close spacing, multiple cropping, drip irrigation, whole-plant harvesting, and other cultural practices will need to be investigated thoroughly to maximize biomass yields while minimizing energy inputs. New harvesting systems and equipment with utilization of a fuel crop as a specific goal will need to be developed, especially for sweet sorghum, for which no reliable large-scale methodology is currently known.

Another crop that is regarded favorably as a potential source of fuel ethanol is corn. Desirable qualities include very high yields, a broad geographical range for cultivation, proven profitability and markets, and photosynthetic efficiency (Battelle Columbus Laboratories 1977*b*). At present, 67% of the corn grain produced is used as domestic animal feed and 30% is exported. A suggested system for production of 10 billion liters of ethanol (roughly enough for a 10% gasohol blend for 20 to 25% of U.S. gasoline) would require an extensive reorganization of the current corn biomass system (Lipinsky 1978). To produce the same quantity of beef, poultry, and pork while retaining current levels of corn acreage and yields, cattle (but not poultry and hogs) would be shifted to corn stover feed and the grain now fed to cattle would be used largely for ethanol production instead. This shift, in turn, would require harvesting and perhaps pretreatment and enrichment of corn stover. Because stover is less digestible than grain, approximately 2 tons of stover would be needed to replace each ton of diverted grain. Widescale use of stover as a replacement cattle feed would preclude its use both as a biomass feedstock for other conversion technologies (Sect. 3.2.1) and as a soil conditioner. In addition, such a shift would entail indirect impacts the same as those of residue collection for fuel conversion (Sect. 5.2.1.2). To be efficient enough to produce an overall net energy gain, this kind of reorganization would have to occur on a gigantic scale. The likelihood of such a scheme becoming acceptable in American agriculture has not yet been evaluated.

3.2.2.2 Other grasses and grains. This category consists of a vast and heterogeneous mix of species. Major groups represented include (1) other members of the grass family (Gramineae), some of which are well known as forage or pasture species whereas others are unconventional candidates for cultivation (e.g., giant reed, *Arundo donax*); (2) the members of the legume family (Leguminosae), such as the clovers, lupines, alfalfas, and kudzu; and (3) a host of miscellaneous groups and species (e.g., cattails, hemp, flax, and sunflower). The category includes wheat, rice, corn, soybeans, barley, and oats but excludes the sugar crops, cotton, tobacco, fruits, and vegetables (Midwest Research Institute 1978). Because all these plants must perform more or less the same physiological functions and thus contain much the same substances (proteins, fats, etc.), the general energy content of plant biomass is given as 16×10^6 J/kg (7000 Btu/lb). Variation is as little as ±5% of this figure for the energy content of hundreds of species (Moss 1977). Plants differ markedly, however, in their tolerances and abilities to use the resources (e.g., light, water and fertilizer) provided by different habitats and hence differ greatly in productivity. Choice of appropriate species for large-scale energy production would be guided by the need to minimize energy inputs while maximizing yields under site-specific conditions.

The extremely diverse nature of the grasses and grains group means that at least some species are potentially likely to provide a biomass crop almost anywhere in the United States where cultivation is plausible. A Midwest Research Institute study (1978) has identified 2.4 million ha (6 million acres) of class I land, 99 million ha (246 million acres) of class II to IV land, and 119 million ha (295 million acres) of class V to VIII land that is "underutilized" (Table 3.5). This designation includes idle cropland and land that is technically tillable although currently devoted to pasture or other low-disturbance uses. Of this total of 222 million ha (547 million acres), about 85% is devoted to livestock production as pasture and range; hence, it is reasonably well established that forage and pasture biomass can be grown on this land successfully although it is largely unsuitable as cropland.

Table 3.5. Underutilized land
[millions of hectares (acres)]

	Class I	Class II—IV	Class V—VIII
Cropland		24 (60)	
Pasture	1.6 (4)	31 (77)	10 (25)
Range	0.4 (1)	39 (97)	106 (263)
Other	0.4 (1)	5 (12)	3 (7)
Total	2.4 (6)	99 (246)	119 (295)

Reprinted by permission of Midwest Research Institute, 1978. *Systems Study of Fuels from Grains and Grasses, Phase 1,* Final Rep., Kansas City, Mo. Prepared for U.S. Department of Energy.

Figure 3.2 shows the distribution of this marginal land. The percentages of land in the Soil Conservation Service capability groups are as follows (Sect. 2.2.1):

1. class II to IV with erosion problems, 28%;
2. class II to IV with wetness problems, 8%;
3. class II to IV with soil problems, 5%;
4. class II to IV with climatic problems, 5%;
5. class V to VIII with erosion problems, 28%;
6. class V to VIII with wetness problems, 2%;
7. class V to VIII with soil problems, 22%;
8. class V to VIII with climatic problems, 3%.

The likely candidates for biomass production are the class II to IV lands with erosion and wetness problems, or about 36% of all underutilized land; this amounts to slightly less than 80 million ha (200 million acres). The major technical problem will be to find combinations of crops and appropriate agricultural practices to ensure substantial long-term productivity, given that most of this land has one or more problems (see Chap. 5).

Cultivation of unconventional crops will require innovations in agricultural practice and technology. For example, cattail is a productive species that might be used to take advantage of available land that is too wet to cultivate without costly drainage operations. Because this plant grows naturally in peat bogs, machinery with flotation capabilities will have to be developed to harvest it. Some type of efficient compressing or dewatering process must also be devised for aquatic plants because the moisture content may be as high as 90%. Where crop production costs cannot be assigned to an established commercial grain crop (i.e., as with crop residue), a farm dedicated solely to biomass production for energy will need to be acutely sensitive to yields and production costs, including those for tillage, fertilization, herbicide and pesticide application, conservation, harvesting, processing, storage, and transport (Midwest Research Institute 1978). Cost constraints alone will eliminate many of the innovative crop production schemes and will narrow the available biomass energy production choices.

The ultimate general methodology for grass, as well as for silvicultural, energy crop production is the "energy plantation"—a systematic approach to the large-scale production of economic fuel from plant matter. Such a proposed plantation is envisioned in one case to be one or more units of a minimum area of 11,500 ha (28,500 acres) (Henry, Frazer, and Vail 1976). For comparison, the average size of all U.S. farms was 214.5 ha (529.7 acres) in the 1970 census, while the average size of farms that grew corn was only 159.7 ha (394.4 acres) (Battelle Columbus Laboratories 1977*b*). The energy plantation would be located close to the fuel-using facility, so that the average fuel transport distance is 17.6 km (11 miles). The farm itself would consist of dense monocultural stands of one or more

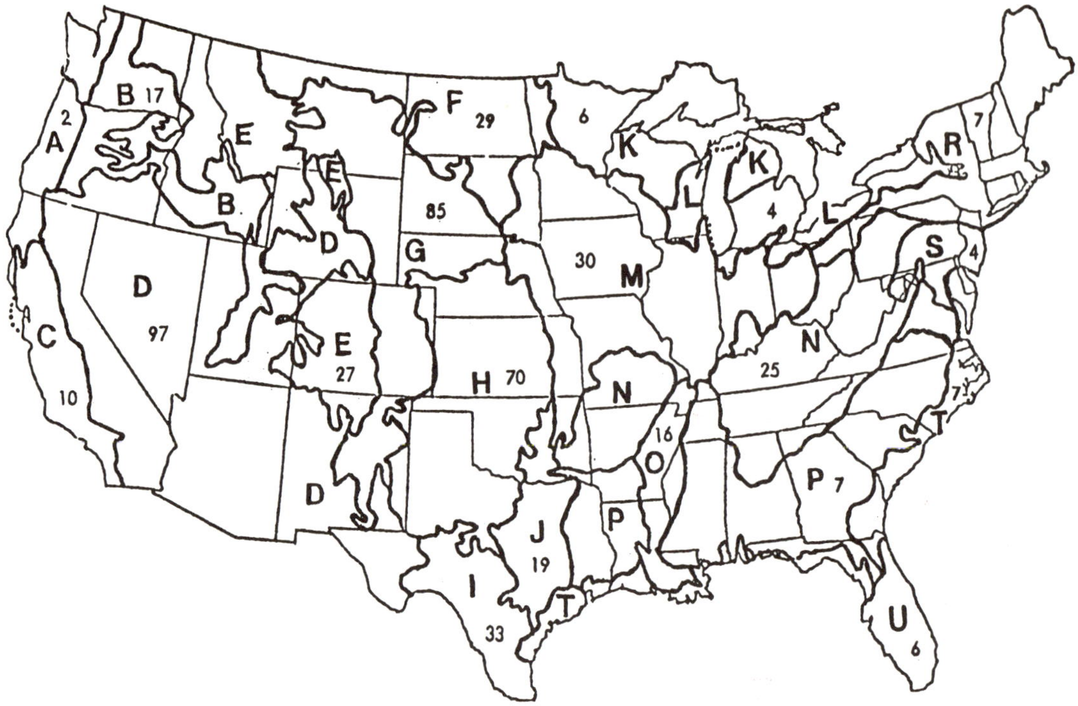

Fig. 3.2. Underutilized acres by land resource region (pasture, range, and other land not in farms) in millions of acres. Source: Midwest Research Institute, Kansas City, Mo., 1978. *Systems Study of Fuels from Grains and Grasses, Phase 1, Final Report*, ALO-3729-1 (based on USDA Conservation Needs Inventory, 1967).

high-yield perennial plant species that are harvested almost year-round at a rate of 17 to 22 metric tons dry weight per hectare. The manpower required for a single minimum-size production unit of 11,500 ha is estimated to be 86 full-time and 48 part-time workers. It is estimated that 40 million hectares (100 million acres) of land with an average productivity of 20 dry metric tons/ha (9 dry tons/acre) could produce about 10.5 EJ (10 quads) of useful fuel per year, or about 14% of the 1974 national annual consumption [77 EJ (73 quads)], under optimal conditions.

Other schemes for energy plantations vary in characteristics such as size—for example, from as small as 1600 to 3200 ha (4000 to 8000 acres) (U.S. Energy Research and Development Administration 1977) to as large as 58,000 ha (144,000 acres) (Stanford Research Institute 1974). Most schemes, however, share the following features:

1. They will be large relative to conventional agriculture.
2. They will grow perennial crops when possible, utilizing intensive management practices and advanced harvesting systems.
3. They will rely on sprouting for reestablishment when possible.
4. They will be established on sites where precipitation is at least 50 to 64 cm (20 to 25 in.) per year, slope is less than 30%, and land is currently used as forest or pasture/range and is producing forage as opposed to field crops (Mitre Corporation 1977).

According to Stanford Research Institute (1974), the general objective of such a plantation is "to produce the greatest amount of biomass possible per unit of time and space, at the lowest possible cost, and with a minimum of energy expenditure."

3.2.2.3 Hydrocarbon crops. The foregoing discussions of sugar and grass crops deal with plants that are only feasible to cultivate in areas receiving at least 50 cm (20 in.) of natural precipitation per year. A vast area of the United States is, however, arid or semiarid and receives less than this amount. Total vegetative growth in desert or semidesert regions is very low. For these areas, interest is less in total production than in the production of plants that manufacture special compounds under conditions of drought stress. These species are hoped to lend themselves to the development of "petrochemical plantation" agriculture (Nielsen et al. 1977). The majority of the world's plants store their energy as carbohydrates (sugars, starches, and cellulose). About 2000 plants (Calvin 1977), however, store energy as special hydrocarbon compounds, the most familiar example of which is *Hevea brasiliensis*, the rubber tree (Calvin 1978), which has long provided latex for the manufacture of natural rubber. Another plant, a member of the Composite family (Compositae) known as guayule (*Parthenium argentatum*), grows naturally in the deserts of the southwestern United States. It also produces a high-molecular-weight hydrocarbon that is usable as a source of rubber. Wild plants of this species are currently being harvested in Mexico for their latex and may very soon become a commercial crop in Texas (Calvin 1978).

Rubber, however useful, is not a fuel; but other plants produce oils, especially the families Euphorbiaceae and Asclepiadaceae. *Euphorbia lathyris*, an annual, has been found to produce a minimum of 10 bbl of crude oil per acre (~ 3930 liters/ha) annually. This figure is likely to be increased substantially by genetic selection processes (Calvin 1978). The caloric value of *E. lathyris* hydrocarbons is estimated at 39×10^6 J/kg (17,000 Btu/lb). Other *Euphorbia* species are trees that can be tapped, like rubber trees, without whole-plant harvesting and that are already known to grow well in southern California. Another plant that is currently the focus of efforts at commercial cultivation is jojoba (*Simmondsia chinensis*), a desert shrub whose oil could not only serve as a source of energy but also apparently surpasses all other known lubricants in its ability to withstand high temperatures without loss of internal stability (Brooks 1978). Present demand for jojoba oil is inhibited by low supply, high price, and the considerable risk factors of establishing plantations. This situation is true of hydrocarbon-producing plants in general. It is also true, however, that the "gasoline trees" may provide an ecologically appropriate way to harvest arid lands without the need to commit scarce water resources for irrigation.

3.3 Livestock Wastes

Compared with crop residues and with plant biomass in general, animal fecal matter is much more troublesome environmentally. A large fraction of the manure produced annually must be treated as waste because of its accumulation at large-scale feeding operation sites that are too remote for economical transportation to fertilizer markets. Research on the use of manure as a feedstock for fuel recovery technologies is thus justifiable in terms of waste management and pollution abatement as well as energy resource terms.

Anaerobic digestion appears to be the most promising conversion technology from both economic and agricultural standpoints (Varani and Burford 1977; Ifeadi and Brown 1975) (see Sect. 5.3). Anaerobic digester technology could be implemented on two levels: as small-scale, on-farm facilities that could contribute to energy self-sufficiency of individual farm units and as commercial-level installations that use inputs from large-capacity feedlots and generate high-quality gas for pipeline transmission or for local use. At present, costs for successful operation favor large-scale installations that involve thousands of cattle and that can afford the necessary high-level capital investment and full-time operators (Thornton 1978; Hashimoto and Chen 1978). On-farm digesters are not generally considered economically feasible at current capital cost and maintenance/management requirements (Thornton 1978). Where pollution control is of primary concern, however, the cost of methane generation could potentially be reduced to a negligible increment in an overall waste management scheme and might even decrease its costs.

3.4 Freshwater Systems

The proposals to use freshwater aquatic biomass for energy production involve (1) microalgal systems, (2) systems using vascular plants such as water hyacinth, and

(3) integrated systems. Except for the possibility of harvesting the nuisance growth of waterweeds, all proposed schemes involve new production of aquatic biomass.

3.4.1 Microalgal Systems

The concept of algal biomass production for conversion to methane originated from the development of algal waste treatment systems at the Sanitary Engineering Research Laboratory, University of California, Berkeley. Microalgae are essential in sewage oxidation ponds; they are highly economical oxygen producers for liquid waste oxidation, and substantial quantities of algal biomass can be produced for which there is no use without expensive processing (e.g., for food or feed). Although production of microalgae may be as much as 87 metric tons/ha-year (39 tons/acre-year), no large-scale systems exist for producing algal biomass because of the lack of cost-effective harvesting methods. This obstacle remains the major one to energy production from algal biomass. Harvesting and processing costs need to be reduced by about one order of magnitude to allow microalgal biomass production solely for fuels (CSO International, Inc. 1978).

Intensive production of microalgae requires shallow ponds designed for water retention times of several days. Mixing should require minimal energy, ideally accomplished with assistance from wind. Existing methods for harvesting algal biomass include centrifugation, flocculation-sedimentation, coagulation-flotation, sand filtration, and microstraining. Drawbacks to these methods include high energy requirements, high cost of coagulation chemicals, and poor suitability for digestion of chemically flocculated biomass. Microstraining uses fine mesh (26 μm) in a rotating drum and water backwash to remove algae from the fabric. This method is applicable only to filamentous algae and would thus require algal species control, which has so far proven difficult to achieve (Benemann et al. 1977).

Two harvesting methods that might become applicable to algal harvest are microstraining and nonchemical sedimentation in isolated separation basins. This latter process relies on the property of autoflocculation, in which microalgae will sometimes spontaneously undergo flocculation and sedimentation. This process is the subject of further research to establish its reliability and to determine the factors that control it (Benemann et al. 1978).

An economic study of large-scale microalgae production (CSO International, Inc. 1978) determined the assumptions necessary to reduce costs to the point of economic feasibility. The assumptions include (1) cultivation of bioflocculating types of algae, (2) a net yield of 20 tons/acre-year, (3) a site with a mild climate and a high insolation rate (land costs not included), (4) use of brackish, saline, or fresh water, (5) low-cost materials and construction techniques, (6) use of combustion flue gas for carbon dioxide enrichment of ponds, (7) recycle of nutrients from digester to pond (or economic credit for use of residue for field fertilization), and (8) harvesting by a pond isolation process. This system was not presented as currently feasible but rather as an exploration of what was required to approach feasibility. The study noted that very large systems (e.g., 100 sq miles) appeared unnecessary because maximum economies of scale for capital investment were reached with modules of about 3 sq miles. The study concluded that the greatest near-term potential was for (1) producing algal biomass as a by-product of sewage treatment or (2) producing microalgae for food, feed, or aquaculture.

3.4.2 Vascular Plant Systems

Emersed plants, those that grow at or above the water's surface, can use atmospheric carbon dioxide. Because they are not limited by the quantity of dissolved carbon in a water body, aquatic vascular plants are thought to have higher growth rates than microalgae. The vascular plants that have been considered for biomass production include water hyacinth (*Eichhornia crassipes*), duckweed (including various species of *Lemna* and *Spirodela*), alligator weed (*Alternanthora philoxeroides*), and *Hydrilla verticillata*. Of these, most attention has been directed toward water hyacinth.

Water hyacinth can have prolific growth rates of 135 dry metric tons/ha-year (60 dry tons/acre-year) or higher (Lecuyer and Marten 1976; Wolverton and McDonald 1979) and can also supply oxygen to and remove nutrients from wastewater. These features and the problems caused by nuisance growth have motivated research on water hyacinth, including investigation of its potential for producing biogas through anaerobic digestion.

The National Aeronautics and Space Administration has studied water purification by using water hyacinth near Bay St. Louis, Mississippi (Wolverton and McDonald 1979). When growing in sewage lagoons, the plants have been shown to remove nutrients and heavy metals such as cadmium, mercury, and nickel. Water hyacinths were harvested by using conveyer systems that deposited cut plants into a boat and an enlarged clamshell bucket operated by a dragline. Harvested water hyacinths were anaerobically digested, producing 0.35 to 0.41 m^3 (12 to 14.5 scf) of biogas per dry kilogram of water hyacinth. The gas contained about 60% methane. Wolverton and McDonald (1979) concluded, based on their observed annual productivity of 154 metric tons/ha in a fertile sewage lagoon, that 1 ha of water hyacinths could produce enough biomass to generate approximately 58,400 m^3 (2.3 million scf) of biogas.

Lecuyer and Marten (1976) analyzed a hyacinth conversion system sized to produce 550,000 m^3/day (200 million scf/day) of pipeline gas. The system, which requires 80,130 ha (198,000 acres) of serpentine canals surroundings a central processing facility, includes 344 growing units, each with two 38-km canals and a collection/feed station wherein harvested material is chopped and pumped as a slurry to the digester. Water is recirculated through the canals. Biogas from the digesters undergoes removal of carbon dioxide to upgrade the gas to pipeline quality. The economic analysis concluded that the project was economically "near-viable" but was "politically questionable due to the large acreage requirements."

Another possibility would be to harvest the existing natural growth of water hyacinth and digest the material to biogas (Wolverton and McDonald 1976). Currently, the U.S. Army Corps of Engineers attempts to control water hyacinth in waterways with mechanical, chemical, and biological methods. This system has produced limited success despite high costs. Although mechanical harvesting of water hyacinth would be expensive, this control method would remove nutrients from the aquatic ecosystem. Neither chemical nor biological control methods can accomplish nutrient removal; upon decomposition the nutrients again become available for additional growth of waterweeds or algae. Although the economics of harvesting the existing growth of water hyacinth have not been explored, it is possible that the costs of harvesting and bioconversion could exceed the value of the biogas produced. However, if digesters were built to use biomass from water hyacinth sewage treatment lagoons, the digesters could accept the additional biomass from weed control operations (Wolverton 1979). The combined economic value of weed control and methane production might make the scheme economically feasible.

3.4.3 Integrated Systems

For both vascular plants and microalgae, systems have been proposed in which the use of biomass for energy is considered to be part of a large, integrated system. This proposal is motivated in part by the poor economics of producing biomass solely for energy and the need to derive higher income from the biomass. Integrated systems may possibly accomplish several tasks (e.g., wastewater treatment, food, energy production) with a lower overall cost than the total costs of several separate processes to accomplish these same tasks.

Oswald et al. (1977) described a spectrum of potential integrated microalgal systems (Table 3.6) that varied in size from 2.6 to 260 km^2 (1 to 100 sq miles) and were envisioned as accomplishing wastewater reclamation (or perhaps disposal of brackish water) in addition to producing fuel or fertilizer. The systems designed primarily for wastewater treatment would conserve large amounts of energy required for conventional wastewater treatment. These systems could perform conventional wastewater treatment in oxidation ponds or advanced wastewater treatment capable of nutrient removal and at the same time produce modest amounts of methane and digester residue having potential value as fertilizer. In the more arid climates, biomass production could be combined with evaporation of irrigation return water, which has high concentrations of dissolved solids. This combi-

Table 3.6. Microalgal biomass production for bioconversion systems

Approximate size range (sq mile)	System	Location	Water source	Nutrients C, N, and P	Products
1	Oxidation ponds	Near urban area	Liquid wastes	Liquid wastes	Methane, fertilizer, O_2
1—10	Advanced wastewater treatment	Near urban areas	Liquid wastes	CO_2, liquid wastes	Methane, fertilizer, nutrient removal
3—30	Complete wastewater treatment	Near urban areas	Liquid wastes	CO_2, liquid wastes, N_2	Methane, fertilizer, evaporative wastewater disposal
5—50	Agricultural fertilizer energy	Rural areas	Drainage irrigation	CO_2, N_2, mineral fertilizer	Methane, nitrogen fertilizer, drainage water disposal
10—100	Nutrient integration	Waste lands	Any locally available source	CO_2 recycled nutrients	Methane

Source: Oswald, W. M., J. R. Benemann, and B. L. Koopman, 1977. "Production of Biomass from Freshwater Aquatic Systems — Concepts of Large-Scale Bioconversion Systems Using Microalgae," Table 1, p. 61, in *Proc. Fuels from Biomass Symposium*, COO/4225-1, U.S. Department of Energy, Washington, D.C.

nation would involve culturing and harvesting algal species tolerant of high salinity. Large-scale algal biomass systems might also be integrated into agricultural production as sites of nitrogen fixation by blue-green algae. However, such systems might require supplemental sources of carbon dioxide, perhaps supplied by flue gases from power generation. Supplemental carbon dioxide is needed because sewage contains only limited amounts of carbon and diffusion of atmospheric carbon dioxide into water is slow. Further research is required to demonstrate the technological and economic feasibility of producing fuel and fertilizer by large-scale culture of algal biomass.

Vascular plants are capable of treating wastewater in lagoons or wetland systems, and a few communities are using or experimenting with such systems (Tourbier and Pierson 1976). Sewage lagoons that use water hyacinths are being adopted or considered in communities in the southern United States. Experience with Brookhaven National Laboratory's experimental system (Small 1978) indicates that a marsh-pond sewage treatment system continues to operate in winter months, although colder temperatures decrease the efficiency of wastewater treatment slightly. These biological systems can achieve considerable energy conservation in wastewater treatment, and in some systems it may be feasible to harvest and use biomass for energy. For example, consultants to the California Water Control Board have proposed a 3800-m^3/day (10^6 gal/day) pilot waste treatment system for the city of San Diego by using water hyacinth ponds. Some energy for plant operation would be derived from anaerobic digestion of fluid extracted from harvested biomass (Martin 1979).

Hillman and Culley (1978) proposed an integrated system involving a duckweed/dairy farm (Fig. 3.3), based on research at Louisiana State University and elsewhere. Duckweed is grown in five 2-acre lagoons, harvested daily by a skimmer, dewatered on a conveyor belt/strainer, spray washed, and mixed with dairy cow feed. Manure from the 100-cow herd is anaerobically digested to produce methane, and the digester residue is pumped into the duckweed lagoons via a heat exchanger connected with the manure collection pit. During the eight-month growing season the lagoons would produce 304 kg/day (670 lb/day dry weight), which would supply 60% of the herd's protein requirement. Any excess production of duckweed could be solar dried or ensiled for feeding during winter months. The lagoons would produce a total of 82 metric tons/year (90 dry tons/year), representing a projected productivity of about 20 metric tons/ha-year (9 dry tons/acre-year).

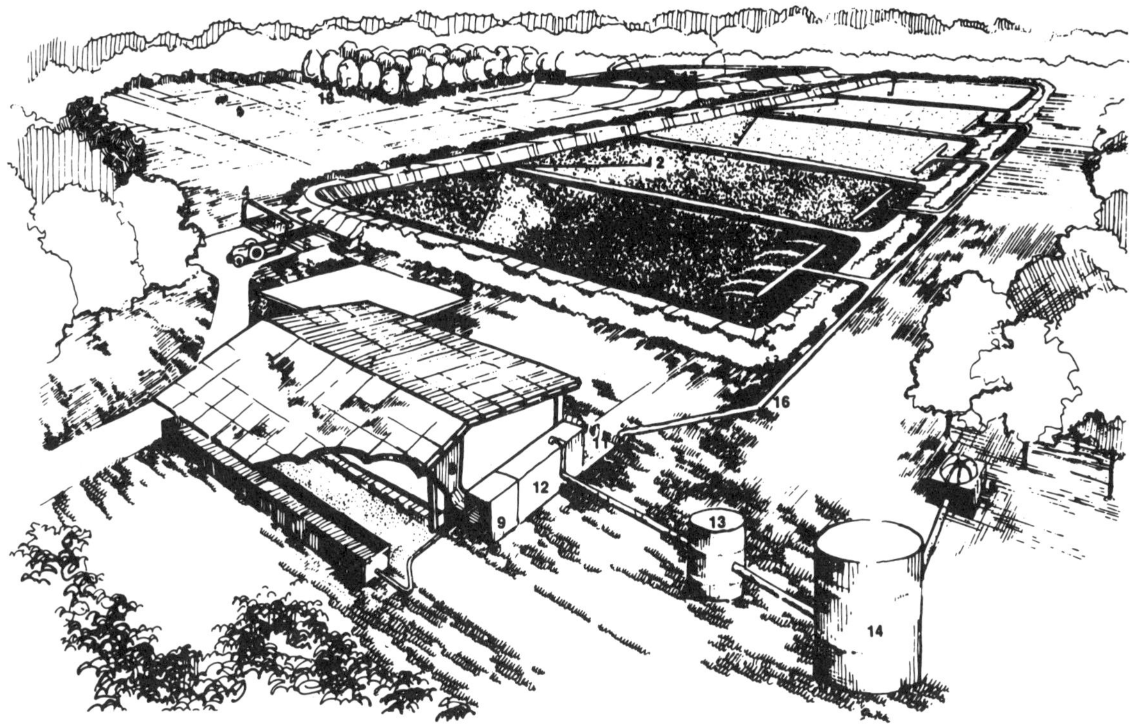

1-3: lagoons, skimmer, and harvesting canal
4-6: washer, conveyor, and washwater basin
7-10: feedlot with solar collector, air duct, heat storage, and drying tunnel
11: waste pit
12-15: fermentation unit, gas scrubber, methane storage, and electric generator
16: sludge to lagoons via heat exchanger in waste pit
17: aquaculture system
18: land application

Fig. 3.3. The duckweed/dairy farm. Source: Hillman, W. S., and D. D. Culley, Jr., 1978. "The Uses of Duckweed," *Am. Sci.* **66:** 442–51, Fig. 2. Used with permission of the publisher.

3.5 Marine Systems

3.5.1 Coastal Systems

A great variety of seaweeds are amenable to mass culture techniques. Among those that could be cultivated in a coastal farm are *Chondrus, Gelidium, Porphyra, Neogardhiella, Undaria, Hypnea,* and *Gracilaria* (Ryther et al. 1977). Many of these species and others are now used as sources of hydrocolloids and food, and all are theoretically convertible to biogas by anaerobic digestion. One method of seaweed culture envisioned would consist of coastal marine production facilities located on land. Long, shallow troughs would hold the seaweed, and municipal wastewater would provide the necessary nutrients. The seaweed culture would effectively act as a wastewater treatment facility in addition to supplying biomass. Alternatively, deep nutrient-rich ocean water could be pumped to the farm to fertilize algal growth. Experimental studies with *Gracilans foliifera* show encouraging yields of about 100 metric tons/ha-year (Ryther et al. 1977). Marine algae could also be grown within net enclosures in natural embayments or behind breakwaters and fertilized with municipal wastes or deep water artificially upwelled from offshore.

The use of coastal marine biomass farms for energy production, however, has many severe drawbacks. Coastal and estuarine areas are extremely valuable biologically, economically, and aesthetically. Competition for coastal regions with present and future users would be intense, and locating large (>100-km^2) marine biomass farms along the coast would probably not be justifiable. In addition, the use of marine algae in land-based coastal farms is often contemplated primarily as a wastewater treatment–mariculture system with energy production only secondary. Ryther et al. (1977) consider it doubtful that such coastal systems could be efficient solely as energy production facilities. For these reasons and the fact that most effort to date has focused on culture of the giant kelp, the method of marine biomass cultivation that will be discussed here is the open-ocean kelp farm.

3.5.2 Oceanic Systems

The greatest potential for the production of large quantities of marine biomass lies in the culturing of kelp on large floating oceanic substrates. Technology is very speculative and it will probably be decades, if ever, before it achieves commercialization. As currently envisioned, each farm would be of modular design and cover an area of about 40,470 ha (100,000 acres). Giant kelp (*Macrocystis pyrifera*) would be grown attached by their holdfast to a grid 18 to 30 m below the water's surface. Nutrient-rich deep water would be pumped up from 100- to 500-m depths by wave-actuated pumps to provide an optimum growing medium at the surface. Periodic harvesting of the top 1.5 m of the kelp beds would provide feedstock for shore-based anaerobic digesters, which would convert the kelp to methane and usable by-products. The conversion of biomass to methane is described in Chap. 4. Estimated productivity ranges between 20 and 250 metric tons/ha-year (dry weight). It is contemplated that various forms of fish mariculture in and around the kelp farm could take advantage of increased primary productivity and shelter. The mariculture subsystem is essentially a means of adding to the cost effectiveness of the project and should not greatly influence the primary function of the farm.

3.5.2.1 The oceanic kelp farm. The concept of an ocean kelp farm producing biomass for conversion to biogas is only about ten years old, and the technology is necessarily in the developmental stages. Preliminary research has included studies on kelp growth in nutrient-rich upwelled water, the behavior of cool upwelled water released at the ocean's surface, kelp transplants onto artificial substrates, and theoretical studies on substrate design and the harnessing of environmental energy (primarily wave energy) (Wilcox et al. 1976; North 1978). A preliminary design for a large-scale commercial ocean kelp farm has been developed; although there will undoubtedly be changes in the final design, the basic parameters enable a general description of the farm and possible environmental impacts. The most comprehensive description of the proposed kelp farm is contained in a series of documents by Wilcox et al. (1976) and was developed, under contract to ERDA, under the auspices of the U.S. Naval Undersea Center, San Diego, California. The following description is based on that series of documents. Current concept development is being coordinated by the General Electric Company, Marine Biomass Program, and is primarily funded by the Gas Research Institute and the Department of Energy. A test farm about 20 m in diameter was deployed off the coast of California in the fall of 1978, but results have not yet been reported.

The basic concept for a single kelp farm is for each large [total area, about 40,470 ha (100,000 acres)] production facility to be composed of many small interconnected modules, each with subsurface grids of synthetic line serving as the actual kelp substrate. Wave-actuated pumps attached to corner support buoys would draw water through vertical upwelling pipes extending 100 to 500 m below the ocean surface and would distribute the cool, nutrient-rich water to the kelp plants as fertilizer. The farm itself would probably not be anchored in deep water but would be dynamically positioned and tensioned by diesel-powered positioning buoys at the corners of the modules. Anchoring may be feasible in shallow water (<1000-m) applications. Conventional harvesting vessels would cut the top portion of the canopy and transport the chopped kelp to onshore processing plants. Expected energy production for a 40,470-ha farm is difficult to predict but might range around 2.7×10^{16} J/year [equivalent to the yearly output of an 800-MW(e) power plant running at 100% load]. A summary of the proposed kelp farm as described by Wilcox et al. (1976) is given in Table 3.7 to provide the reader with an impression of such systems.

Table 3.7. Generalized technical summary of the ocean kelp farm

Systems	Farm size[a] 405 km^2 (40,470 ha) (100,000 acres — 12.5 miles/side)[b]
	Amount/number
Buoys	
Corner support	5,000
Catenary	4.4 million
Wave pump	20,000
Lines	
Kelp support	134,000 km
Heavy tension	4,600 km
Catenary buoy	134,000 km
Total	272,600 km
Upwelling pipe	
Vertical	5,000
Horizontal distribution	6,600 km
Positioning diesels	5,000
Upwelling water	13,000 m^3/s

[a]This table is designed to provide the reader with an impression of the magnitude of proposed ocean kelp farms. The eventual system will most likely differ substantially from the above description based on ongoing research and development. All values are gross approximations only and can vary greatly depending on underlying assumptions.

[b]Values taken from Wilcox, H. A., et al., 1976. *Ocean Food and Energy Farm Project Subtasks 1–7*, ERDA/USN/1027-76, Energy Research and Development Administration, Washington, D.C.

3.5.2.2 Kelp support system. The support system would provide a surface for attachment of the kelp, maintain the kelp at the proper depth, and concatenate other farm components such as upwelling pumps, the water distribution piping, and the positioning system.

Proper substrate depth is a compromise among stability (protection from wave damage), biological requirements of the kelp, kelp productivity, and cost. A substrate depth of between 20 and 30 m appears to maximize these parameters. Modular design of the ocean kelp farm is contemplated because of unavoidable scaling problems for a unitary system of necessary size. Modules afford advantages of size independence, assembly ease, and the ability to localize damage. Kelp mooring lines will be of light synthetic material spaced about 3 m apart. Heavy synthetic mooring lines for tension will be used at the perimeter of the modules. These flexible modules will probably be free-floating [anchoring in deep water is prohibitively expensive (Wentworth et al. 1978)]. Diesel-driven buoys at the corners of each module will serve dual functions of maintaining both the shape of the nonrigid module (tensioning) and its position on the surface of the ocean. A schematic diagram of the module is shown in Fig. 3.4. Six of these triangular modules, about 300 m on a side, would be joined to form a hexagonal covering of about 28 ha. About 1400 of these hexagonal modules would comprise the entire 40,470-ha farm.

Kelp plants could be attached to the grid in various ways. Plants may be affixed to the subsystem by allowing sporophylls to shed spores over "attachment elements" in the laboratory and subsequently transporting these elements to the farm and attaching them to the grid support system. An alternative method would be to attach mature plants to the substrate at large intervals and allow natural culturing (settlement) of the remaining portions of the substrate. Preliminary data from the test farm indicate that this last desirable option may be quite viable.

3.5.2.3 Fertilization. Open-ocean surface waters are typically low in the nutrients necessary for kelp growth. Cold deep waters at depths of 100 to 500 m, however, have high concentrations of nitrogen and phosphorus, and coastal upwelling naturally provides nutrient-rich water to existing kelp beds. Researchers contemplate that water from these intermediate depths can be brought to the surface by harvesting environmental energy. Wave-actuated pumps at each corner support buoy will be attached to 100- to 500-m-long, vertical pipes extending downward to the desired depth. Nutrient-rich water pumped to the surface would be distributed to the kelp through a system of surface pipes. Municipal sludge and wastewater and/or commercial fertilizers could be used to augment fertilization. The amount, type, mode of dispersal, appropriateness, and necessity for such augmentation have yet to be determined.

3.5.2.4 Positioning of the kelp farm. Approximately 170 MW (230,000 hp) is estimated to be necessary to position the kelp farm in the currents expected off the coast of California if anchoring proves unfeasible. This will be provided by numerous (>5000) small diesel engines located in corner buoys that will maintain tension in the flexible modules in addition to positioning the farm. The energy expense for this function will be substantial.

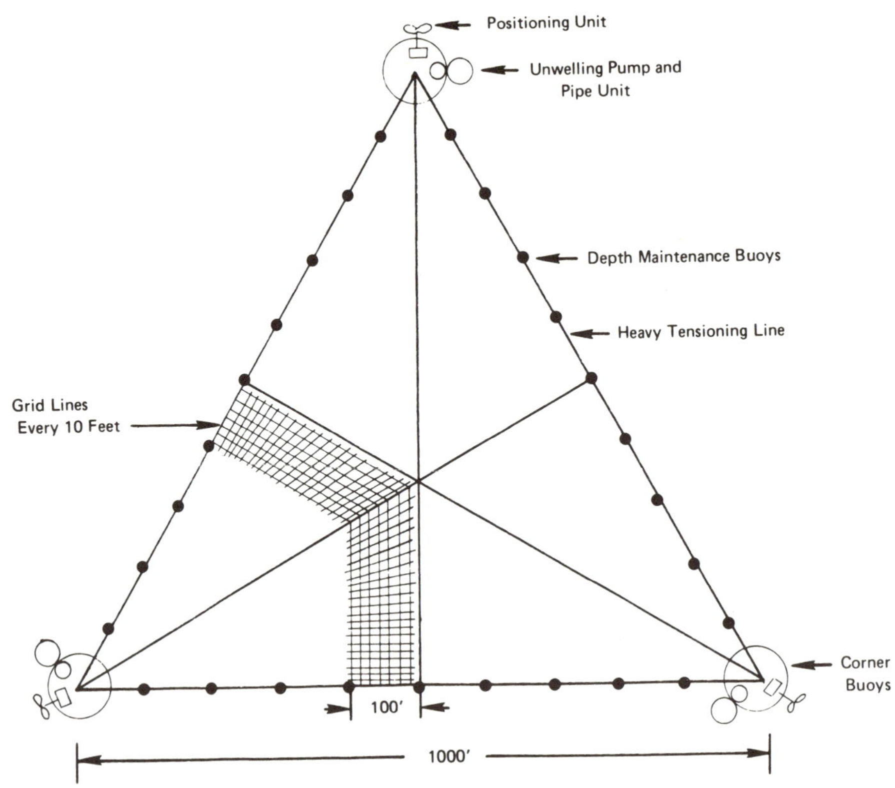

Fig. 3.4. Ocean farm substrate—module schematic diagram, top view. Many modules would be linked to form the farm. This is only one of many possible configurations. Source: Wilcox, H. A., et al., 1976. *Ocean Food and Energy Farm Project*, vols. 1-7, ERDA/USN/1027-76, prepared for U.S. Energy Research and Development Administration by U.S. Naval Undersea Center, San Diego, Calif.

However, an alternative may be to introduce the farm into an appropriate oceanic gyre and allow it to cycle indefinitely, with the engines primarily used to correct position within the gyre and to maintain grid tension.

3.5.2.5 Life support system. Farm supervision and maintenance will be partially provided by personnel located and supported onsite. The support system is envisioned to house about 50 people on a 100-m-square floating platform (barge), which will have facilities for housing, storage, maintenance, communication, and propulsion and will include a helipad. It, like the kelp modules themselves, will be free floating and will use diesel power to maintain position in relation to the ocean farm.

3.5.2.6 Kelp farm deployment. Deployment of the ocean kelp farm is restricted by a variety of factors. High water temperatures damage or kill kelp, while cold temperatures inhibit growth. Environmental temperatures of between 18 and 25°C appear to be appropriate. The farm site can extend somewhat south of the 25°C thermal contour because upwelling water will cool surface waters to some extent. Water clarity is desirable to maximize kelp growth. Storm tracts are obviously to be avoided, but appropriate sea states will be necessary to provide wave energy for upwelling. The farm should be located a distance from shore to minimize both transportation costs and interference with coastal shipping and must also be located in water that is nutrient rich at 100 to 500 m depth. Interference with existing commercial fisheries and recreational areas of the proposed site must be avoided.

References

Alexander, A. G., 1977. *Production of Sugarcane and Tropical Grasses as a Renewable Energy Source,* First Quarterly Report, ORO/5422-1, U.S. Department of Energy, Washington, D.C.

Bascom, W., 1964. *Waves and Beaches,* Doubleday, New York.

Battelle Columbus Laboratories, Columbus, Ohio, 1978. *Potential Benefits and Technology Development Costs for Sugar Crops as an Energy Source.*

———, 1976. *Systems Study of Fuels from Sugarcane, Sweet Sorghum, and Sugar Beets,* vol. 2, *Agricultural Considerations,* BMI-1957.

———, 1977*a*. *Systems Study of Fuels from Sugarcane, Sweet Sorghum, and Sugar Beets,* vol. 1, *Comprehensive Evaluation, Task 77,* BMI-1957.

———, 1977*b*. *Systems Studies of Fuels from Sugarcane, Sweet Sorghum, Sugar Beets and Corn,* vol. 5, *Comprehensive Evaluation of Corn,* BMI-1957.

Benemann, J. R., 1978. *Biofuels: A Survey,* ER 746-SR, Electric Power Research Institute, Palo Alto, Calif.

Benemann, J. R., et al., 1977. *Species Control in Large-Scale Algal Biomass Production, Final Report,* ERDA SAN/7405-77/1, U.S. Energy Research and Development Administration, Washington, D.C.

Benemann, J. R., et al., 1978. "Fuels from Microalgae Biomass," pp. 891–946 in *Proc. 2d Ann. Symposium on Fuels from Biomass, Troy, N.Y., June 20–22, 1978,* U.S. Department of Energy, Washington, D.C.

Biomass Energy Institute, Inc., 1978. *Biomass Energy Success Stories, A Portfolio Illustrating Current Economic Uses of Renewable Biomass Energy,* NTIS HCP/T0285-01, prepared for the U.S. Depatment of Energy, Washington, D.C.

Brooks, Witt, 1978. "Jojoba—A North American Desert Shrub: Its Ecology, Possible Commercialization, and Potential for Introduction into Other Arid Regions," *J. Arid Environ.* **1:** 227–36.

Burwell, C. C., 1978. "Solar Biomass Energy: An Overview of U.S. Potential," *Science* **199:** 1041–48.

Calvin, G., and M. Calvin, 1978. *The Only Source of Energy,* TID-4500-R66, prepared for the U.S. Department of Energy by Lawrence Berkeley Laboratories, Berkeley, Calif.

Calvin, M., 1978. "Green Factories," (Priestly Medal address), *Chem. Eng. News* **56:** 30–36.

———, 1977. "The Sunny Side of the Future," *Chem. Technol.* **7**: 352–63.

CSO International, Inc., Concord, Calif., 1978. *Cost Analysis of Microalgae Biomass Systems, Final Report: Design Assumptions and Engineering Plans for a 100-square-mile System.*

Donovan, N.S., 1977. *Wave Energy and Beach Response for Three Southern Rhode Island Beaches,* M.S. Thesis, University of Rhode Island.

Duke, J. M., and M. J. Fudali, 1976. *Report on the Pulp and Paper Industry's Changing Fuel Mix,* American Paper Institute, New York.

Hashimoto, A. G., and Y. R. Chen, 1978. *Anaerobic Fermentation of Livestock and Crop Residues, Quarterly Progress Report,* prepared for the U.S. Department of Energy.

Henry, J. F., M. D. Frazer, and C. W. Vail, 1976. "The Energy Plantation: Design, Operation, and Economic Potential," in *Thermal Uses and Properties of Carbohydrates and Lignins,* F. Shatizaden, K. V. Sankanen, and D. A. Tillman, eds., Academic Press, New York.

Hewett, C. E., 1978. *The Availability of Wood for a 50-MW Wood-Fired Power Plant in Northern Vermont,* DSD No. 114, Research Program on Technology and Public Policy, Dartmouth College, Hanover, N.H.

Hillman, W. S., and D. D. Culley, Jr., 1978. "The Uses of Duckweed," *Am. Sci.* **66:** 442–51.

Howlett, K., and A. Gamache, 1977*a*. *Silvicultural Biomass Farms,* vol. 6, *Forest and Mill Residues as Potential Sources of Biomass.* Mitre Corp. Tech. Rep. 7347, McLean, Va.

Howlett, K., and A. Gamache, 1977*b*. *Silvicultural Biomass Farms,* vol. 2, *Biomass Potential of Short Rotation Farms,* Mitre Corp. Tech. Rep. 7347, McLean, Va.

Ifeadi, C. N., and J. B. Brown, Jr., 1975. "Technologies Suitable for Recovery of Energy from Livestock Manure," in *Energy, Agriculture and Waste Management,* W. J. Jewell, ed., Ann Arbor Science Publishers, Ann Arbor, Mich.

Inman, R. E., 1977. *Silvicultural Biomass Farms,* vol. 1, *Summary,* Mitre Corp. Tech. Rep. 7347, McLean, Va.

Iowa State University, Ames, Iowa, 1977. *Energy Relationships for a 320-Acre Iowa Farm,* Rep. PB-275274, prepared for U.S. Department of Commerce.

Jamison, R. L., 1977. "Trees as a Renewable Energy Resource," in *Clean Fuels from Biomass and Wastes,* Institute of Gas Technology, Chicago, Ill.

King, C. A. M., 1965. *An Introduction to Oceanography,* McGraw-Hill, New York.

Koch, Peter, and D. W. McKenzie, 1976. "Machine to Harvest Slash, Brush and Thinnings for Fuel and Fiber—A Concept," *J. For.* **74:** 809-912.

Lecuyer, R. P., and J. H. Marten, 1976. "An Economic Assessment of Fuel Gas from Water Hyacinths," pp. 267-86, in *Symposium Papers: Clean Fuels from Biomass, Sewage, Urban Refuse, and Agricultural Wastes, Orlando, Fla., Jan. 27-30, 1976,* Institute of Gas Technology, Chicago, Ill.

Lipinsky, E. S., 1978. "Fuels from Biomass: Integration with Food and Materials Systems," *Science* **199:** 644-51.

Martin, Cecil, California State Water Resources Commission, Sacramento, Calif., Apr. 24, 1979. Telephone conservation with R. D. Roop, ORNL.

Midwest Research Institute, Kansas City, Mo., 1978. *Systems Study of Fuels from Grains and Grasses, Phase 1, Final Report,* ALO-3729-1.

Miles, T. R., 1977. "Logistics of Energy Resources and Residues," pp. 225-48 in *Fuels and Energy from Renewable Resources,* D. A. Tillman, K. V. Sarkunen, and L. L. Anderson, eds., Academic Press, New York.

Mitre Corporation, 1977. *Systems Descriptions and Engineering Costs for Solar and Related Technologies,* vol. 9, *Biomass Fuels Production and Conversion Systems,* Tech. Rep. 7485, McLean, Va.

Moss, D., 1977. "An Agronomist's View of Biomass Production," in *Biomass—A Cash Crop for the Future?, Conference on the Production of Biomass from Grains, Crop Residues, Forages, and Grasses for Conversion to Fuels and Chemicals, Kansas City, Mo., Mar. 2-3, 1977,* MRI and Battelle Columbus Laboratories for U.S. Energy Research and Development Admin., Columbus, Ohio.

Nielsen, P. E., et al., 1977. "Plant Crops as Sources of Fuel and Hydrocarbon-like Materials," *Science* **198:** 942-44.

(NCFES) North Central Forest Experiment Station, 1978. *Forest Residues Energy Program, Final Report,* TID-28416, U.S. Department of Energy, Washington, D.C.

North, W. J., and C. L. Hubbs, 1968. *Utilization of Kelp-Bed Resources in Southern California,* Calif. Dept. Fish and Game Bull. 139, Sacramento, Calif.

North, W. J., 1978. "A General Description of Macrocystis," in *Handbook of Solar Biosources,* vol. 1, *Basic Principles,* C. C. Black, and A. Mitsui, eds., CRC Press, Inc., Cleveland, Ohio.

Oswald, W. M., J. R. Benemann, and B. L. Koopman, 1977. "Production of Biomass from Freshwater Aquatic Systems—Concepts of Large-Scale Bioconversion Systems Using Microalgae," p. 61 in *Proc. Fuels from Biomass Symposium,* John Pfeffer and J. J. Stukel, eds., COO/4225-1, U.S. Department of Energy, Washington, D.C.

Ryther, J. H., et al., 1977. "Cultivation of Seaweeds as a Biomass Source for Energy" pp. 83-113 in *Proc. Fuels from Biomass Symposium,* John Pfeffer and J. J. Stukel, eds., COO/4225-1, U.S. Department of Energy by University of Illinois, Urbana.

Shephard, F. P., 1973. *Submarine Geology,* Harper and Row, New York.

Small, M. W., 1978. "Low Energy Wastewater Treatment," pp. 383-85 in *Proc. of the 1978 Conference on Technology for Energy Conservation, Albuquerque, N. Mex., Jan. 24-27, 1978,* Information Transfer, Inc., Washington, D.C.

Stanford Research Institute, Menlo Park, Calif., 1974. *Effective Utilization of Solar Energy to Produce Clean Fuel,* PB 233956.

———, 1977. *An Evaluation of the Use of Agricultural Residues as an Energy Feedstock—A Ten-Site Study,* vol. 1, *Summary and General Information.*

Steinbeck, K., 1978. "Short-Rotation Hardwood Forestry in the Southeast," pp. 175-84 in *Proc. 2d Annual Symposium on Fuels from Biomass, Troy, N.Y., June 20-22, 1978,* U.S. Department of Energy, Washington, D.C.

Sverdrup, H. V., M. W. Johnson, and R. H. Fleming, 1942. *The Oceans—Their Physics,*

Chemistry, and General Biology, Prentice-Hall, Inc., Englewood Cliffs, N.J.

Thornton, Ted, 1978. *An Assessment of Anaerobic Digestion in U.S. Agriculture,* ESCS-06, Economic, Statistical, and Cooperative Services, U.S. Department of Agriculture, Washington, D.C.

Tillman, D. A., 1978. *Wood as an Energy Source,* Academic Press, New York.

Tinberg, C., and A. D. McElroy, 1978. "The Economics of Growing Grasses for Energy," pp. 859-72 in *Proc. 2d Annual Symposium on Fuels from Biomass, Troy, N.Y., June 20-22, 1978,* Rensselaer Polytechnic Institute, Troy, N.Y.

Tourbier, J., and R. W. Pierson, Jr., 1976. *Biological Control of Water Pollution,* University of Pennsylvania Press, Philadelphia.

(USDA) U.S. Department of Agriculture, 1973. *The Outlook for Timber in the U.S., 1970,* Forest Service Res. Rep. 10, Washington, D.C.

U.S. Energy Research and Development Administration, Washington, D.C., 1977. *Solar Program Assessment: Environmental Factors, Fuels from Biomass,* ERDA 77-47/7.

Varani, F. T., and J. J. Burford, Jr., 1977. "Conversion of Feedlot Wastes into Pipeline Gas." pp. 87-104, in *Fuels from Waste,* L. L. Anderson and D. A. Tillman, eds., Academic Press, New York.

Wayne, C. J., 1976. "The Effects of Sea and Marsh Grass on Wave Energy," *Coastal Res. Notes* **4**(7): 6-8.

Wentworth, R. L., et al., 1978. "Cost Analysis of Growing and Harvesting Aquatic Biomass," *2d Annual Symposium on Fuels from Biomass, Rensselaer Polytechnic Institute, Troy, N.Y., June 20-21, 1978,* pp. 873-79.

Wilcox, H. A., 1976. "Ocean Farming," pp. 255-96 in *Proc. Conf. on Capturing the Sun Through Bioconversion, Washington, D.C., Mar. 10-12, 1976,* Bio-Energy Council, Washington, D.C.

Wilcox, H. A., et al., 1976. *Ocean Food and Energy Farm Project,* vols. 1-7, ERDA/USN/1027-76, prepared for U.S. Energy Research and Development Admin. by U.S. Naval Undersea Center, San Diego, Calif.

Wolverton, B. C., NASA, Bay St. Louis, La., Apr. 18, 1979. Telephone conversation with R. D. Roop, ORNL.

Wolverton, B. C., and R. C. McDonald, 1976. "Water Hyacinths (*Eichhornia crassipes*), a Renewable Source of Energy," pp. 249-52 in *Proc. Conf. on Capturing the Sun Through Bioconversion, Washington, D.C., Mar. 10-12, 1976,* Bio-Energy Council, Washington, D.C.

———, 1979. "The Water Hyacinth: From Prolific Pest to Potential Provider," *Ambio* **8**(1): 2-9.

Zerbe, John, 1976. "Agriculture and Forestry Wastes: Comments," pp. 144-56 in *Proc. Conf. on Capturing the Sun Through Bioconversion, Washington, D.C., Mar. 10-12, 1976,* Bio-Energy Council, Washington, D.C.

4

Overview of conversion

Biomass conversion involves essentially two considerations: (1) the economic and technical feasibility of producing energy from biomass and (2) the consequences of that production. This section describes biomass materials, conversion processes, end products, effluents, and full-scale conversion more in the context of consequences than of feasibility. Because much of the technology is still tentative, materials and processes are discussed primarily in terms of potential—potential for producing both energy and environmental impact. In most cases, very little information is available on full-scale operating processes and their emissions, whereas much is available on possible schemes and promising ideas for converting biomass. Thus, the coverage here, rather than being comprehensive, provides a basis for discussion of effects.

4.1 Description of Biomass Materials

The primary source of energy in biomass is reduced carbon or carbohydrates, which appear as sugar, starch, or cellulose. The important sugar-bearing biomass species are sugarcane, sweet sorghum, and sugar beets, and an important source of starch is corn kernels. Sugar and starch as energy materials require cultivation of specific crops. Cellulose, however, is the chief structural material in all plants and appears widely in all vegetation, especially in crop waste. Wheat, rye, oat, rice, and soybean straws; corn stover; sugarcane bagasse; all animal manures; and wood and wood wastes are highly cellulosic. Cellulose comprises almost one-third of the weight of all trees, grasses, and straws (McKetta and Cunningham 1978).

4.1.1 Carbohydrates

Carbohydrates are the principal product of photosynthesis, the process whereby green plants use light energy to convert atmospheric carbon dioxide into reduced forms of carbon (carbon combined with hydrogen) in compounds like sucrose ($C_{12}H_{22}O_{11}$), starch ($C_6H_{10}O_5$), and cellulose. The energy that is recoverable from biomass depends, at least biochemically, not only on the carbohydrate content but also on the form of the carbohydrate. Calvin (1976) showed that theoretically sugars can be converted directly to alcohol by microorganisms with almost no energy loss in the conversion:

$$C_6H_{12}O_6 \xrightarrow[\text{yeast or bacteria}]{} 2C_2H_5OH + 2CO_2\ ,$$

$$\underset{(673\ \text{kcal})}{2818\ \text{kJ}} \longrightarrow \underset{(655\ \text{kcal})}{2742\ \text{kJ}}\ .$$

About 1.3 kg of sugar is needed to produce 1 liter of alcohol [22,190 kJ (5300 kcal)] (Calvin 1976). Starch is also easily fermented after it has been treated gently with an enzyme-containing material. However, cellulose requires more extensive treatment. Cellulose consists of individual glucopyranose units that are linked repeatedly to form a linear polymer of about 50 to 5000 units (Ott et al. 1954). Thus, a pretreatment step such as acid or enzymatic hydrolysis is necessary to saccharify the cellulose, that is, to convert the long-chain polymeric unit to simple sugars and thereby make it accessible to fermentation (Sect. 4.5.4.1).

4.1.2 Lignocellulosic Materials

The term *lignocellulosic* refers to the structural portion of plant matter that primarily contains cellulose in close physical and chemical association with lignin. The cellulose constitutes the essential part of woody cell walls and, along with hemicellulose, amounts to 95% of wood. The other 5% is made up of gums, fats, resins, waxes, sugars, oils, starches, alkaloids, and tannins.

The structure and physical properties of wood are determined primarily by its three principal lignocellulosic components: cellulose, hemicellulose, and lignin. The relative amounts of these components vary as shown in Table 4.1. Two major differences are apparent between softwoods and hardwoods: the softwoods have a higher lignin content and contain mannan, the polymeric form of mannose, in the hemicellulose portion, whereas hardwoods contain xylan, the polymeric form of xylose, in place of mannan in the same fraction. The term *hardwood* refers to the botanical broad-leaved group of trees, whereas softwoods (or conifers) have needle- or scale-like leaves and are generally evergreen. As a combustion fuel the two types of wood differ considerably in that the softwoods burn more readily and rapidly with a more intense flame. This difference results primarily from the higher volatile resin and oil (pitch) content of softwood, which, pound for pound, has almost twice the heating value of wood itself (Table 4.17, Sect. 4.2.1) (Johnson and Auth 1951).

Like cellulose, hemicellulose is a carbohydrate, but it differs from cellulose in that it is amorphous, more readily hydrolyzed, and composed of pentoses rather than hexoses. Lignin, which comprises 18 to 28% of wood, is the noncarbohydrate portion of the cell wall. The lignin is an amorphous polymer composed of substituted phenyl propane units. Hardwood and softwood lignins differ; not only is the total amount of lignin different, the molecular species vary as well. Not readily degraded by microorganisms, lignin protects the cellulosic portion of plant materials from microbial degradation and must be destroyed or removed prior to fermentation. Wood tars, which have a high phenolic and polycyclic aromatic hydrocarbon (PAH) content, are thought to be derived from the lignin portion of wood, which is highly aromatic (Wolfhagen 1979).

The conspicuous absence of aromatic compounds in the pyrolysis of lignin-free wood cellulose (Table 4.2) attests to the role of lignin as the precursor of aromatics in wood conversion products. This is not to suggest that lignin is a sole precursor of aromatics. High concentrations of polycyclic aromatic compounds are found in carbon black produced from natural gas (Lee and Hites 1976), which has little, if any, aromatic content. Conversion of biomass to energy and the outputs therefrom depend on many factors, including (1) the kind of feedstock; (2) the moisture content; and, most importantly, (3) the conditions under which it is converted—namely, feed rate, size of particles, air-to-feedstock ratio, temperature of conversion, and configuration of the firing area.

Analyses of several types of wood wastes are given in Table 4.3. Two types of analyses, proximate and ultimate, are performed on fuel materials. Both are performed under standardized conditions, but the results can vary considerably, depending on the laboratory performing the analyses (Table 4.3). Proximate analysis determines moisture, volatile-matter, fixed-carbon, and ash content of the biomass, whereas ultimate analysis provides information on elemental composition (carbon, hydrogen, nitrogen, oxygen, and sulfur). Fixed carbon is the nonvolatile fraction other than ash and is determined by difference—that is, by subtracting both the weight of evolved volatile matter and the final ash weight from the initial weight of the dry sample. The fixed-carbon fraction would correspond to the nonvolatile burnable part of charcoal remaining after pyrolyzing wood. Because wood waste averages about 50%

Table 4.1. Chemical composition (%) of some North American species of trees (extractive-free wood basis)

Species	Analysis				Estimation										Gross composition		
	Ash	Lignin	Acetyl	Uronic acid anhydride	Glucan	Galactan	Mannan	Araban	Xylan	Cellulose	Non-cellulosic glucan?	Glucomannan (acetate)	Arabinogalactan	4-*O*-methyl-glucurono (arabino) xylan (acetate)	Cellulose	Hemicelluloses	Lignin
Trembling aspen *Populus tremuloides*	0.2	16.3	3.4	3.3	57.3	0.8	2.3	0.4	16.0	53	3	4	1	23	53	31	16
White elm *Ulmus americana*	0.3	23.6	3.9	3.6	53.2	0.9	2.4	0.6	11.5	49	2	4	2	19	49	27	24
Beech *Fegus grandifolia*	0.4	22.1	3.9	4.8	47.5	1.2	2.1	0.5	17.5	42	4	4	2	25	42	36	22
White birch *Betula papyrifera*	0.2	18.9	4.4	4.6	44.7	0.6	1.5	0.5	24.6	41	2	3	1	34	41	40	19
Yellow birch *Betula lutea*	0.3	21.3	3.3	4.2	46.7	0.9	3.6	0.6	20.1	40	3	7	1	28	40	39	21
Red maple *Acer rubrum*	0.2	24.0	3.8	3.5	46.6	0.6	3.5	0.5	17.3	41	2	7	1	25	41	35	24
Sugar maple *Acer saccharum*	0.3	22.7	2.9	4.4	51.7		2.3	0.8	14.8			4	1	22			
Hardwoods, average										45	3	5	1	25	45	34	21
Balsam fir *Abies balsamea*	0.2	29.4	1.5	3.4	46.8	1.0	12.4	0.5	4.8	44	0	18	1	8	44	27	29
Eastern white cedar *Thuja occidentalis*	0.2	30.7	1.1	4.2	45.2	1.5	8.3	1.3	7.5	44	0	11	2	12	44	25	31
Eastern hemlock *Tsuga canadensis*	0.2	32.5	1.7	3.3	45.3	1.2	11.2	0.6	4.0	42	0	17	1	7	42	26	33
Jack pine *Pinus banksiana*	0.2	28.6	1.2	3.9	45.6	1.4	10.6	1.4	7.1	41	0	16	2	12	41	30	29
White spruce *Picea glauca*	0.3	27.1	1.3	3.6	46.5	1.2	11.6	1.6	6.8	44	0	17	2	10	44	29	27
Tamarack *Larix laricina*	0.2	28.6	1.5	2.9	46.1	2.3	13.1	1.0	4.3	43	0	18	3	7	43	28	29
Softwoods, average										43	0	16	2	9	43	28	29

Source: Brooks, R. E., W. D. Bellamy, and T. M. Su, 1978. Table XIV, p. 78, in *Bioconversion of Plant Biomass to Ethanol*, COO/4147-4, U.S. Department of Energy, Washington, D.C.

Table 4.2. Pyrolysis products of holocellulose from loblolly pine bark with and without catalysts

No.	Compound	Total yield (%)		
		Neat	5% $ZnCl_2$	5% NaOH
1—4	Low-molecular-weight hydrocarbons	0.40	0.43	1.80
5	Methanol	1.12	1.72	0.65
6	Acetaldehyde	0.21	0.15	0.34
7	Acrylaldehyde	0.07	0.05	0.09
8	Furan	0.52	0.64	0.47
9	Acetic acid	1.39	0.54	1.04
10	Diacetyl	0.23	0.24	0.31
11	1-Hydroxy-2-propanone	0.06	0.02	0.22
12	2-Furaldehyde	0.51	0.77	0.06
	Carbon monoxide	5.28	5.80	8.98
	Carbon dioxide	11.04	11.81	20.42
	Water	37.25	42.51	37.22
	Char	20.17	34.83	28.04

Reprinted by permission of Fang, P., and G. D. McGinnis, 1976. "Flash Pyrolysis of Holocellulose from Loblolly Pine Bark," p. 41 in *Thermal Uses and Properties of Carbohydrates and Lignins*, F. Shafigadih, K. V. Sarkanen, and D. A. Tillman, eds., Academic Press, New York.

moisture, 40% volatiles, and 10% fixed carbon, with ash less than 1% of the dry wood, most of the combustible carbon is in volatile gases. If these gases are not properly ignited during combustion, they can constitute a major source of air emissions (Sect. 4.4.2.1).

Although ash is only a small percentage of dry wood, it will constitute the major solid waste from almost any wood conversion process. According to Ellis (1962), surprisingly little interest has been shown in precisely identifying the components of wood ash, which contains essentially all the mineral matter in the wood (Table 4.4). The broad range of compositions given in Table 4.4 attests to the generally accepted notion that the amounts of mineral components in wood do not follow a fixed pattern; rather, they vary widely not only with species but also with the available mineral content of the soils in which the tree was grown (Ellis 1962). Calcium, potassium, and sodium are the dominant elements in wood ash, as compared with silicon, aluminum, and iron in coal ash (Braunstein et al. 1977). However, the ash content of wood bark can be considerably higher than that of the heartwood (Sect. 4.1.2.2 and Table 4.6).

4.1.2.1 Hogged fuel. The term *hogged fuel* is usually applied to wood wastes such as slabs, bark, trimmings, and edgings that have been machine hogged, or put through a mill containing rapidly rotating knives or swing hammers, to reduce them to a uniformly small size. However, the term is also applied to wood

Table 4.3. Analyses of wood-waste fuels[a]

	Kind of fuel						
	California redwood		Western hemlock		Douglas fir		Pine sawdust
Laboratory	B.M.	F.E.	B.M.	F.E.	B.M.	F.E.	B.M.
Moisture, as received, %	50.4	44.3	57.9	57.9	35.9	41.4	
Moisture, air dried, %	7.3		7.3		6.5		6.3
Proximate analysis, % of dry fuel							
Volatile matter	82.5	82.2	74.2	78.9	82.0	82.2	79.4
Fixed carbon	17.3	17.5	23.6	19.5	17.2	17.5	20.1
Ash	0.2	0.3	2.2	1.6	0.8	0.3	0.5
Ultimate analysis, % of dry fuel							
Hydrogen	5.9	5.6	5.8	5.9	6.3	5.8	6.3
Carbon	53.5	53.4	50.4	50.1	52.3	52.5	51.8
Nitrogen	0.1	0.1	0.1	0.1	0.1	0.1	0.1
Oxygen	40.3	40.5	41.4	42.2	40.5	41.2	41.3
Sulfur	0	0.1	0.1	0.1	0	0.1	0
Ash	0.2	0.3	2.2	1.6	0.8	0.3	0.5
Total	100.0	100.0	100.0	100.0		1000.0	
Btu/lb (dry)	9920.0	8840.0	8620.0	8290.0	9050.0	8730.0	9130.0

[a]B.M. = Bureau of Mines; F.E. = Fuel Engineering Company, New York, N.Y.

Reprinted by permission of Kreisinger, H., 1939. "Combustion of Wood-Waste Fuels," *Mech. Eng.* 61: 115—19, Table 1.

Table 4.4. Composition of wood ash

Element	Assumed form	Typical quantity (% of total ash)
Calcium	CaO	30—60
Potassium	K_2O	10—30
Sodium	Na_2O	2—15
Magnesium	MgO	5—10
Iron	Fe_2O_3	1—2
Silicon	SiO_2	2—7
Phosphorus	P_2O_5	5—15
Sulfur	SO_3	1—5

Note: These figures were determined under conditions that ensure complete combustion and the retention of all ash generated. The percentage composition was calculated by assuming that the elements are in the stated chemical form; the actual chemical forms may be different.

Reprinted by permission of Shelton, J., 1978. Table 3-2, p. 23, in *The Woodburner's Encyclopedia*, Vermont Crossroads Press, Waitsfield, Vt. Data interpreted from L. E. Wise, ed., *Wood Chemistry*, Reinhold Publ. Corp., New York, 1944, p. 435.

Table 4.5. pH values for cold-water extractions of plant residues

Residue type	Value
Maple shavings	5.22[a]
Red oak sawdust	4.53
Red oak bark	4.87
Walnut bark	6.03
White oak chips	4.52
White oak bark	5.25
Cottonwood bark	5.58
Corncobs	5.17
Cornstalks	7.49
Hickory shavings	5.63
Mixed hardwood shavings	5.00
Cottonwood shavings	7.38
Sunflower seed coats	6.90
Aspen fiber (commercial)	

[a]Each value is an average of five measurements. Reprinted by permission of Chow, P., C. S. Walters, and J. K. Guiher, 1971. "pH Measurements for Pressure-Refined Plant Fiber Residues," *For. Prod. J.* **21**(12): 51—52, Table 1.

shavings, sawdust, bits of bark, chips, and other residue from the lumber industry. The quantity of waste in lumbering is large, representing about one-half of all timber cut. For this reason lumbering wastes are an important source of fuel energy. When stored outdoors, hogged fuel loses 10 to 25% of its heating value during the first six months of storage because of escaping volatiles (Johnson and Auth 1951). No information was found describing the nature of these volatilized materials, but they should not necessarily be considered inconsequential because they are undoubtedly carbonaceous and may contain environmentally significant compounds. Additionally, wood waste stored outdoors can give rise to an acidic runoff (Table 4.5). In wet climates rain and snowfall will produce runoff, whereas in dry climates periodic wetting is required to prevent accidental ignition of the dry wood waste.

4.1.2.2 Bark. Like hogged fuel, bark from lumbering waste is an important source of energy. It has been reported that 650 to 1050 kg of wet bark are recovered per metric ton (1300 to 2100 lb/ton) of unbleached pulp from hemlock or white fir logs (Johnson and Auth 1951). The amount of bark is strongly dependent on the season in which the logs are harvested. For some woods, as much as 75% more bark will be obtained from logs cut in the fall and winter than from those cut in the spring when the sap has begun to flow (Johnson and Auth 1951). The moisture content of bark may range from 45 to 80%. The wide difference is caused by handling procedures, the higher value being the result of both river driving and wet hydraulic debarking. A high water content has important consequences in that particulate emissions are increased (Sect. 4.4.2.1) and heating value of the fuel is decreased (Sect. 4.2.1.1).

Table 4.6 gives analyses of wood bark from various trees. The inherent ash content of bark, which is between 0.5 and 5%, is five to ten times that of whole wood. Additionally, much soil and sand mineral matter is incorporated into the bark during dragging, causing the production of much larger quantities of ash during combustion.

The nitrogen content of wastewood is listed as 0.1% in Table 4.3. However, an in-depth analysis of all parts of various trees indicates that the nitrogen concentrations in bark are three to nine times higher than those in wood (Young and Guinn 1966), which is important in evaluating both emissions and possible environmental impacts (Sects. 4.4.1.1 and 6.1.1).

Table 4.6. Typical analyses of wood bark[a]

	Douglas fir	Cedar	Lodgepole pine	White fir	Hemlock	Spruce	Redwood
Saltwater storage	No	No	No	Yes	Yes	Yes	Yes
Proximate analysis							
Volatile matter, %	72.9	73.0	73.5	74.3	72.5	69.6	72.6
Fixed carbon, %	25.9	21.9	26.0	24.0	25.8	26.6	27.0
Ash, %	1.3	5.1	0.5	1.7	1.7	3.8	0.5
Sulfur, %	Trace	Trace	Trace	Trace	0.03	0.1	Trace
Btu/lb (dry)	9570	8610	9310	8810	9090	8740	8350
Ultimate analysis							
Carbon, %	56.2	51.0	55.0	52.2	53.7	51.8	51.9
Hydrogen, %	5.9	5.7	5.8	5.8	5.7	5.7	5.1
O_2 and N_2 (difference)	36.7	38.2	38.7	40.3	38.9	38.6	42.6
Ash analysis							
Silica, SiO_2	13.9	44.0	12.2	1.7	1.5	32.0	14.3
Iron as Fe_2O_3	4.4	7.2	5.6	3.2	3.6	6.4	3.5
Titanium oxide, TiO_2	0.4	0.8	0.3	0.0	0.0	0.8	0.3
Alumina, Al_2O_3	8.7	13.8	8.7	3.2	1.7	11.0	4.0
Manganese, Mn_3O_4	0.3	0.1	0.2	3.9	3.3	1.5	0.1
Calcium oxide, CaO	51.4	13.9	48.0	60.8	58.2	25.3	6.0
Magnesium oxide, MgO	3.2	3.1	1.8	3.0	4.4	4.1	6.6
Sodium oxide, Na_2O	5.3	6.6	8.4	10.4	9.1	10.4	25.0
Sulfate as SO_3	2.9	3.1	3.3	3.0	3.7	2.1	7.4
Chloride, Cl	0.4	0.7	0.4	0.4	1.4	Trace	18.4
Carbonate, as CO_2	7.3	5.2	9.1	11.5	11.8	7.0	14.0
Unaccounted for	1.8	1.5	2.0	1.1	1.3	0.6	0.4

[a]Compiled by Babcock & Wilcox Co. Research and Development Department.

Reprinted by permission of Johnson, A. J., and G. H. Auth, 1951. *Fuels and Combustion Handbook*, McGraw-Hill, New York.

Bark or other high-moisture wood fuels are commonly fired with a small percentage (10 to 15%) of a fossil fuel such as oil or coal to improve the combustibility of the wet fuel. Mixing dirty bark with coal, especially bark stored in saltwater, can reduce the fusion temperature of the coal ash, causing clinker and slagging problems (Johnson and Auth 1951). Additionally, the nature of emitted particulates and bottom ash would then depend not only on the wood or bark fired but also on the coal and incorporated foreign materials.

4.1.3 Grain Crop Residues

The compositions of common crop residues are given in Table 4.7. Compared with wood wastes, dry crop wastes have a very high ash content, of which as much as 85% may be silica. Both rice hulls and rice straw have exceptionally high silica compositions, and rice hulls can exceed some woods in lignin content. All the residues have relatively equal amounts of cellulose and hemicellulose.

4.1.4 Sugar Crop Residues

Of all possible biomass feedstocks for conversion to energy, sugar crop residues are the leading contenders because of their favorable composition, high yield, low unit cost, and availability in large quantity at the point of use (i.e., for firing boilers in a fermentation plant). For example, bagasse, the refuse that remains after the extraction of sugar from sugarcane, amounts to 20 wt % of the sugarcane. The structure, combustion behavior, and analysis of bagasse are not unlike those of wood. Normally, bagasse is composed of 35 to 45% fiber and 7 to 10% sucrose as combustible material, with 45 to 55% moisture. The dry material contains 45% carbon, 6% hydrogen, 46% oxygen, and 1.5 to 3% ash. Like wood, dry bagasse possesses 18,608 to 20,934 kJ/kg (8000 to 9000 Btu/lb) of energy, but the high moisture content reduces the heating value to between 9304 and 13,956 kJ/kg (4000 and 6000 Btu/lb) (Sect. 4.2.1.1). Thus, it is not uncommon to cofire bagasse with fossil fuels such as oil or gas. The intrinsic biomass ash content is

Table 4.7. Compositions of crop residues, dry matter basis (%)

Residue	Ash	Silica	Cellulose	Hemicellulose	Lignin
Corncob		0.4	34.2—43.3	37.3—49.8	5.9—8.1
Corn stover	9.72	7.0—8.3	29.9—41.0	23.4—33.4	6.1—9.7
Grain sorghum stover		3.7	33.1—38.0	20.0—21.7	7.8—9.3
Oat straw	9.35	2.5—4.0	35.2—42.8	22.4—26.1	8.0—11.5
Rice hull		21.4	34.2—43.0	7.0—21.0	14.7—17.2
Rice straw		4.9—16.5	33.2—42.6	15.0	6.0—7.0
Soybean stover	7.65		37.0—44.9	13.8—22.4	9.1—16.8
Wheat straw		3.5	40.9—49.9	24.5—27.4	9.3—12.5

Source: Smith, R. J., J. A. Miranoiuski, and R. L. Vetter, 1977. *Energy Relationships for a 320-Acre Iowa Farm,* ERDA/USDA-19464/76/FR-3, Iowa State University, Ames.

relatively low, but, as with dirty bark, soil and silt may be incorporated into the cane during harvesting, causing an increase in residue ash and slagging problems during combustion. Despite the moisture, bagasse burns readily and rapidly on an incandescent bed, requiring feed rates often in excess of that required for burning wood. Like wood, the more finely the bagasse is divided, the easier it is to burn. When a modern bagasse furnace is used, efficiencies greater than 60% can be achieved, even with material containing 50% moisture (Johnson and Auth 1951). The composition and fuel value of sugar beet tops and sweet sorghum fiber may not be too different from that of sugarcane bagasse, considering the characteristics of the total plants (Sect. 4.1.6 and Table 4.10).

4.1.5 Corn

Grain corn, or corn kernels, has received serious consideration as a feedstock material, both for fermentation to fuel ethanol (Lipinsky et al. 1977*b*) and for direct combustion as an additive to coal (Cioffi 1978). An analysis of the corn plant is given in Table 4.8. As it grows, the grain constitutes 53% of the plant on a dry-weight basis, and starch constitutes about 80% of the grain. The remainder of the grain is the cellulosic hull and the fatty germ tissue. The nongrain portion of the corn plant (stover) is primarily cellulosic and is combustible in the same manner as sugarcane bagasse (Sect. 4.1.4).

As a supplemental fuel to be burned with coal in a pulverized-coal-fired utility boiler, grain corn is similar to bagasse or hardwood bark (Table 4.9). Compared with coal, corn has more than twice the volatile matter, less than one-fourth the fixed carbon, and one-tenth the ash content. However, the Btu value is only two-thirds that of coal. The energy content of a fuel depends on the degree of oxidation of the carbonaceous material in the fuel; the greater

Table 4.8. Compositional analysis of the corn plant

	Dry matter (%)	Residues (%)	Dry basis (%)						
			Crude Protein	Total ash	Hemi-cellulose	Cellulose	Lignin	Fats and oils (est.)	Other solubles[a]
Grain	77		10.9	1.7		2.4[a]		4.7	80.3[b]
Leaf	82	11.7	7.0	13.7	24.8	22.0	5.1	2.4	25.0[c]
Husk	57	13.0	2.8	3.4	36.4	32.9	6.7	0.9	16.9[c]
Cob	56	21.3	2.8	1.4	37.3	34.9	7.4	0.4	15.8[c]
Stalk	34	53.9	3.7	4.7	23.7	34.4	10.5	1.7	21.3[c]

[a]Crude fiber 76.3%.
[b]Starch.
[c]Primarily sucrose.
Source: Lipinsky, E. S., et al., 1977. Table 2, p. 15, in *Systems Study of Fuels from Sugarcane, Sweet Sorghum, and Sugar Beets,* vol. 5, BMI-1975, Battelle Columbus Laboratories, Columbus, Ohio.

Table 4.9. Analysis of corn as a supplemental fuel

	Coal sample C-14461	Corn sample M-30400
Total moisture, %	17.6	17.3
Proximate analysis (dry), %		
Volatile matter	40.3	87.6
Fixed carbon	45.7	11.2
Ash	14.0	1.2
Grindability (ASTM D-409)	67	27
Btu/lb (dry)	12,230	8130
Btu/lb (maf)[a]	14,220	8230
Ultimate analysis (dry), %		
Carbon	68.1	46.5
Hydrogen	4.9	6.5
Nitrogen	1.2	1.6
Sulfur	3.1	0.1
Ash	14.0	1.2
Oxygen (diff.)	8.7	44.1
Total	100.0	100.0

[a]maf = moisture and ash free.

Source: Cioffi, P. L., 1978. Table 2 in *Corn/Coal Fuel Characterization Study*, Report HCP/T3052-01, U.S. Department of Energy, Washington, D.C.

the oxidation, the lower the energy content. Biomass materials have a relatively low energy content compared with that of hydrocarbon fuels because cellulose, sugars, and hemicelluloses (the components of vegetable matter) are partially oxidized forms of carbon. However, the low sulfur, nitrogen, and ash contents give these fuels a favorable environmental advantage. Because of the lower Btu value of corn, 1.5 lb of corn would be required to equal the heat input of 1 lb of coal. Therefore, the feed rate for corn would have to increase by 50% to maintain an equivalent heat input. However, a 10% corn-coal blend would require only a 5% increase in total fuel input rate. The combustion performance of the 10% corn-coal blend would be improved over that of coal in terms of ignitability, stability, and turn-down capability, whereas the emission of sulfur and particulates would be reduced. Also, the quantity of ash would be reduced (Cioffi 1978). Thus, corn as a combustion fuel would have a high environmental acceptability insofar as emissions are concerned. However, the negative socioeconomic impact of using a food directly as fuel could easily overwhelm any other benefits (Chap. 7).

4.1.6 Sugar Crops

The compositions of sugarcane, sugar beets, and sweet sorghum are given in Table 4.10, and their cultivation characteristics are described in Sect. 3.2.2.1. Despite the fundamental species and cultivation differences of the three crops, their proximate analyses are

Table 4.10. Compositions of typical sugar crops[a]

	Sugarcane			Sugar beets			Sweet sorghum	
	Stalks	With tops and leaves[b]	Mill-run bagasse	Roots	Tops	Total	Sugar type	Syrup type
Moisture, %	69	73	49	78	84	81	77	77
Soluble organics, %	15	16	4[c]	15			14	14
Ash, soluble, %		1		1			1.6	1.7
Combustible organics, %	*d*	9	45[e]	*f*	13	*d*	7	7
Ash, insoluble, %		1	2	1	4	3[g]	0.4	0.3
Gross heating value, MJ/metric ton		5045	9530					
Net heating value, MJ/metric ton		3451	7620					
On a moisture-free, sugar-free basis, %								
Cellulose			47.8					
Pentosans			29.2					
Lignin			19.3					
Ash			3.7					

[a]Interview data, Louisana State University.
[b]Estimated from interview data and values for bagasse and solubles.
[c]Absorbed solubles.
[d]Not calculated.
[e]Fiber.
[f]Not used for fuel.
[g]Total ash.

Source: Nathan, R. A., 1978. Table 6.1, p. 61, in *Fuels from Sugar Crops — Systems Study for Sugarcane, Sweet Sorghum, and Sugar Beets*, TID-22781, U.S. Department of Energy, Washington, D.C.

surprisingly similar. As sources of carbohydrate for fermentation to liquid fuels, all three crops are equally important, but other considerations limit their potential for conversion (Lipinsky et al. 1977*a*).

4.1.7 Animal Manures

The conversion of animal wastes, especially cattle feedlot manure, into energy and energy products is considered expedient for extracting energy as well as for preventing the environmental degradation from runoff that accompanies open storage of these wastes (Sect. 2.3.5). The chemical composition of manure from beef cattle is given in Table 4.11. In addition to the constituents listed in the table, manures may contain trace elements in concentrations that are five to ten times greater than in the animal feed. Capar et al. (1978) have analyzed feedlot manure and feedlot diet; their results are presented in Table 4.12. The toxic trace elements, As, Cd, Hg, Pb, and Sb, are five- to tenfold more concentrated in the manure than in the feedlot diet. Further concentration of these elements in digestion sludge by conversion of manures to methane may inhibit use of these waste solids as fertilizers (Sect. 6.3) and may even preclude their use for continuous recycle feeding.

Table 4.11. Chemical composition of fresh manure from beef cattle

Constituent[a]	Wet weight (%)
Water	85.0
Dry solids	15.0
Volatile solids	11.9
Ash	3.1
Carbon	7.5
Nitrogen-organic	0.32
Phosphorus	0.06
Potassium	0.15
Calcium	0.15
Magnesium	0.06
Sodium	0.15
Sulfur	0.05
Iron	0.06

[a]pH = 7.3; density of dry manure = 0.96 kg/liter (60 lb/ft^3); heating value of dry manure = 3429 kcal/kg (6173 Btu/lb).

Source: U.S. Environmental Protection Agency, 1978. *Preliminary Environmental Assessment of Biomass Conversion to Synthetic Fuels.*

Table 4.12. Comparison of elemental levels in animal feeds, animal wastes, and sewage sludge[a]

Elements	FLD/AF	AW/FLD	SS/FLD
Macronutrients			
Ca	0.84	9.6	9.4
Cl	1.2	4.2	2.2
K	0.96	3.4	1.5
Mg	0.82	2.8	3.7
Na	1.3	6.5	5.0
P	0.78	2.5	5.4
Mean	0.98	4.8	4.5
Micronutrients			
Co	1.1	16.4	96
Cr		17.4	1920
Cu	0.33	7.9	481
Fe	2.5	10.3	128
Mn	0.52	9.4	11.4
Mo		>8.6	>5.7
Se		3.0	15.5
Sn		>5.5	>270
V		7.9	71.2
Zn	1.2	5.7	107
Mean	1.1	>9.2	>311
Toxic elements			
As		9.8	143
Cd		6.6	2080
Hg		>5	>860
Pb		10.1	5089
Sb		>5.3	>353
Mean		>7.4	>1705
Nonessential elements			
Al		8.1	28
Ba		6.6	34.5
Br		3.4	4.6
Eu		11.3	308
La		16.5	71.4
Rb		8.4	14.2
Sc		9.4	38.5
Ti		6.8	288
Mean		8.8	98.4

[a]All numbers are ratios of elemental levels (ppm on air-dried weight basis); FLD, feedlot diet; AF, animal feed; AW, animal wastes; SS, sewage sludge.

Source: Capar, S. G., et al., 1978. "Multielement Analysis of Animal Feed, Animal Wastes, and Sewage Sludge," *Environ. Sci. Technol.* **12**(7): 785-90, Table 3. Reprinted by permission of American Chemical Society, Copyright 1978.

4.1.8 Marine Biomass

Two seaweed systems have been proposed for energy production: the giant alga *Macrocystis pyrifera* and red algae such as *Gracilaria* and *Chondrus crispus* (Jackson 1977) (Sects. 3.5.1 and 3.5.2). Algae differ from terrestrial plants by taking all their nutrients from solution through photosynthesizing tissue (Jackson 1977). The part of algae that is important for energy production is the cell wall, which consists of two or three layers of cellulose along with other complex carbohydrates (Fig. 4.1). The harvested plant contains only 12.5% solids, of which about 40% is ash. However, about 45% of the nonash portion is fermentable carbohydrate, which, unlike the starch, cellulose, and simple sugars of most land plants (Sect. 4.1.1), is made up of complex mucilagenous polysaccharides ordinarily classified as vegetable gums and pectins (Hoagland 1915). The ash content of kelp, which can vary widely and with no apparent variation pattern (Hart et al. 1976), is made up of salts—primarily sodium and potassium chlorides, with some sodium sulfate. Sulfur is an important component of kelp (Table 4.13). Amounts average about 1%, with the form about equally divided between organic and inorganic. Although the intrinsic sulfur content of kelp is small, the annual output of a minimum mariculture farm (100,000 acres) can be significant (Sects. 4.4.5 and 6.3.1).

4.1.9 Freshwater Biomass

The primary prospects for fuel from freshwater biomass are microalgae and water hyacinth, although other prolific aquatic "weeds," such as duckweed, have also been considered (Sects. 3.4.1 and 3.4.2). Microalgae are free-floating microscopic plants that derive all their nutrients from water. Of the thousands of species that exist, Table 4.14 lists species that are most abundant and that are

Table 4.13. Percentage of sulfur in dried Pacific Coast kelp

No.	Material used	Total sulfur	Inorganic sulfur	Organic sulfur
1.1	*Macrocystis pyrifera*, leaves	1.25	0.55	0.70
1.2	*Macrocystis pyrifera*, stems	0.82	0.38	0.44
2.1	*Macrocystis pyrifera*, leaves	1.14	0.46	0.68
2.2	*Macrocystis pyrifera*, stems	0.72	0.36	0.36
15.1	*Pelagophycus porra*, leaves	1.03	0.49	0.54
15.2	*Pelagophycus porra*, stems	0.71	0.28	0.43
29.1	*Nereocystis luetkeana*, leaves	1.27	0.82	0.45
29.2	*Nereocystis luetkeana*, stems	0.45	0.14	0.31
42	*Egregia menziesii*	1.17	0.32	0.85
43	*Laminaria andersonii*	2.12	1.07	1.05
44	*Iridaea* spp.	8.16	4.52	3.64

Reprinted by permission of Hoagland, D. R., 1915. "Organic Constituents of Pacific Coast Kelp," *J. Agric. Res.* 4: 39–58, Table 4.

Table 4.14. Algae most abundant and widespread in oxidation ponds

Rank	Genus
1	*Chlorella*
2	*Ankistrodesmus*
3	*Scenedesmus*
4	*Euglena*
5	*Chlamydomonas*
6	*Oscillatoria*
7	*Micractinium*
8	*Golenkinia*

Source: Benemann, J. R., et al., 1977. *Species Control in Large-Scale Algal Biomass Production*, Report SAN/7405-77/1, U.S. Department of Energy, Washington, D.C., Table 1.

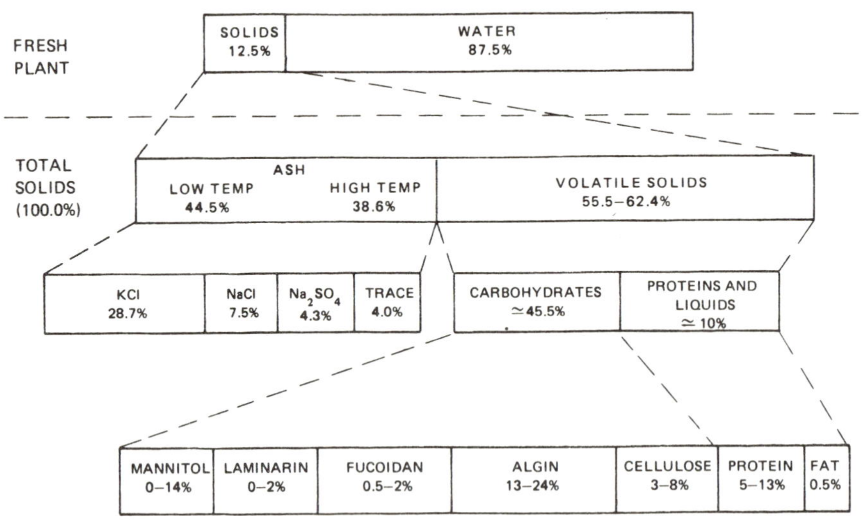

Fig. 4.1. Composition of *Macrocystis*. Source: Hoffman, G. A., and V. Budhraja, 1976. *Ocean Food and Energy Farm Project Subtask No. 6: Systems Analysis*, ERDA/USN/1027-76/1, U.S. Energy Research and Development Administration, Washington, D.C.

considered useful because they provide oxygen in sewage ponds. The fermentable portion of a typical green algal cell is the two-layered cellulose and pectin cell wall and stored starch.

Unlike algae, both duckweed and water hyacinth are rooted vascular plants, although duckweeds are the smallest and simplest of flowering plants (Hillman and Culley 1978). Instead of distinct leaves and stems, the plant body represents a fusion of both. The compositions of several species of duckweeds grown in waste ponds are given in Table 4.15. The distribution of components varies considerably from that of marine biomass (Sect. 4.1.8 and Fig. 4.1).

Water hyacinth is the most commonly considered freshwater aquatic species for conversion to methane. Generally, the plants float in shallow water and are rooted in the muddy sediment. They have slender stalks and an abundance of floating leaves and are often surrounded by clusters of many flowers. The growth and reproductive characteristics have been described by Del Fosse (1977). Table 4.16 gives the composition of water hyacinth, showing variations with growth location. The carbohydrate and cellulose are the most significant components for conversion. As with all fermentation feedstocks, the final disposition of the sulfur and heavy metals will be environmentally significant (Sect. 6.3.1).

4.2 Description of Conversion Processes

The energy potential in biomass can be released by direct combustion or by converting these materials into fuels or chemicals either thermochemically or biochemically. The following sections discuss technologies for combustion and conversion. Although greater detail is given for processes that have advanced beyond the laboratory experimental stage, an attempt has been made to include sufficient data to provide a broad overview of the various processes, products, and potential wastes. Details of full-scale conversion, including facilities, are given in Sect. 4.5.

4.2.1 Combustion

Combustion of biomass is usually described as proceeding through four consecutive steps: (1) evaporation of moisture, (2) distillation of volatiles, (3) burning of volatiles, and (4) combustion of fixed carbon. Step 1 is endothermic, step 2 can be either endothermic or exothermic, and steps 3 and 4 are definitely exothermic, providing the desired heat. The vaporization of moisture absorbs about 2675 kJ/kg (1150 Btu/lb) [2442 kJ/kg (1050 Btu/lb) of moisture evaporated plus 233 kJ/kg (100 Btu/lb) to break the hydroscopic bond between water and wood] (Shelton 1978).

Table 4.15. Approximate compositions of duckweeds grown on various waste-treatment sites

Treatment source	Species	Component (% dry weight)						
		Crude protein	Fat	Fiber	Ash	Ca	K	P
Municipal waste lagoons	*Spirodela oligorhiza*	32.7	6.3	13.5	20.3	1.5	1.8	1.1
Anaerobic swine waste lagoon	*S. oligorhiza*	44.7	4.6	8.3	13.4	1.0	2.8	1.8
Municipal waste lagoon	Mixture of *Lemna minor* and *L. gibba*	39.0						
Dairy cattle waste lagoon	*S. oligorhiza*	37.8	3.8	7.3	12.0	1.3	2.9	1.5
	S. polyrhiza	40.9	6.7	8.7	12.9	2.1	2.1	1.4
	L. gibba (strain G3)	38.5	3.0	9.4	16.4	1.0	4.2	1.6

Reprinted by permission of Hillman, W. S., and D. D. Culley, Jr., 1978, "The Uses of Duckweed," *Am. Sci.* **66**: 442–51, Table 1.

Table 4.16. Mean mineral content of nutrients and other parameters in water hyacinth from various locations (dry weight measures)

Parameter	Amount	Location	Parameter	Amount	Location
Al, %	18.60	Guyana	Mg, %	1.10	Florida
				0.55	Florida
Bo, %	20.00	Florida		1.05	Alabama
				4.30	Guyana
Ca, %	1.00	Florida			
	1.35	Florida	Md, %	12.00	Florida
	3.10	Guyana			
			Mn, ppm	270.00	Florida
Carbohydrate, % total available	7.80	Florida		3,940.00	Alabama
				3,000.00	Guyana
			N, %	1.75	Florida
Cellulose, %	28.00	Florida		2.39	Florida
				2.64	Alabama
Cl, %	18.46	Guyana			
			Na, %	0.41	Florida
Crude protein, %	20.60	Florida		0.34	Alabama
	10.90	Florida			
	12—18	Florida	P, %	0.42	Florida
	13.00	Florida		0.43	Alabama
			ppm	0.54	Florida
Cu, ppm	15.00	Florida		0.43	Alabama
	11.00	Alabama			
			P_2O_5, %	3.20	Guyana
Fe, ppm	3,420.00	Florida			
	250.00	Alabama	S, %	0.33	Florida
	52,000.00	Guyana		0.48	Florida
K, %	4.40	Florida	SO_3, %	4.40	Guyana
	4.45	Florida			
	4.25	Alabama	Zn, ppm	67.00	Florida
	20.30	Guyana		50.00	Alabama
kcal/g	3.80	Florida			
Methylglucose, % in roots	80.00	Dacca			

Reprinted by permission of Del Fosse, E. S., 1977. "Water Hyacinth Biomass Yield Potentials," Table 3, p. 84, in *Clean Fuels from Biomass and Waste Symposium Papers*, Institute of Gas Technology, Chicago, Ill.

Not only is vaporization of moisture the most important heat-absorbing portion of the combustion process, it is the source of increased particulate emissions (Arola 1977), requiring so much heat that biomass containing more than 60% moisture will barely support combustion. Step 2, the distillation step, is also somewhat endothermic, below about 260°C (500°F), although it probably absorbs no more than about 465 kJ/kg (200 Btu/lb) of distilled volatile matter (Kreisinger 1939). The bulk of the distillation process, however, takes place between 282 and 482°C (540 and 900°F) and is exothermic in this temperature range (Shelton 1978). In the presence of air at a temperature of about 593°C (1100°F), the evolved volatiles ignite and burst into flame, producing large quantities of heat. Among others, the most abundant compounds volatilized are carbon monoxide, carbon dioxide, methane, wood alcohol or methanol, formaldehyde, hydrogen, and formic and acetic acids. The heats of combustion for some of these compounds are given in Table 4.17. In the absence of sufficient air or at temperatures below the ignition point, these gases are emitted uncombusted and

Table 4.17. Heat of combustion of selected compounds

Compound	Heat of combustion (Btu/lb)
Carbon	14,540[a]
Hydrogen	61,000[a]
Carbon monoxide	4,380[a]
Methane	24,000[a]
Methanol	9,600[b]
Acetic acid	6,300[b]
Formaldehyde	8,000[b]
Formic acid	2,400[b]
Pitch	15,100[b]
Pressed sawdust	7,300[b]

Sources: (*a*) Johnson, A. J., and G. H. Auth, 1951. Chap. 5 in *Fuels and Combustion Handbook*, McGraw-Hill, New York; and (*b*) Dean, J. A., 1973. *Lange's Handbook of Chemistry*, 11th ed., McGraw-Hill, New York.

constitute a major source of wood-burning chemical pollutants. This is almost invariably the case when wood is burned in open air or in residential fireplaces (Sects. 4.4.1 and 4.5.1.3). In fact, few wood stoves are designed to provide the additional forced air and temperature necessary to ignite and burn the evolved gases. Commercial boilers, on the other hand, are designed to extract all the possible heat from fuel and thus generally provide more complete combustion.

4.2.1.1 Moisture in wood. The effect of moisture content on the heating value of wood is given in Table 4.18. Clearly, the heat required to evaporate the moisture in wood takes a severe toll on the originally available heat. At 50% moisture, which is common for wood as received or undried wood of 20,900 kJ/kg (9000 Btu/lb), only 9000 kJ/kg (3900 Btu/lb) remain to provide heat and maintain a suitable combustion temperature. Lowering the temperature of the fuel bed has several important environmental consequences. A low temperature is conducive to high particulate emissions (Sect. 4.4.2.1) and to incomplete combustion, with its consequent emission of unburned hydrocarbons and combustible gases such as carbon monoxide.

4.2.2 Thermochemical Conversion

Figure 4.2 is a flow diagram of thermochemical processes, their interfaces, and their associated products. As shown, pyrolysis, liquefaction, and gasification are the three primary thermochemical processes through which biomass is converted into useful fuels or chemicals. Other unit processes indicated in Fig. 4.2, such as shift conversion and methanation, provide cleanup and refinement of the product. Char, oil, and low- or medium-Btu gas are products obtained directly from the thermochemical processes. Low- or medium-Btu gas can be converted to methane, methanol, hydrogen, and ammonia. Methanation technology is currently under development, whereas methanol and ammonia synthesis technolo-

Table 4.18. Effect of moisture content on the heating value of wood and wood products

Moisture (%)	Heat value per lb (Btu)	Heat required to evaporate water[a] (Btu)	Net heat remaining (Btu)	Water/ dry wood (lb/lb)
0	9000	0	9000	0.00
10	8100	120	7890	0.11
20	7200	240	6960	0.25
30	6300	360	5940	0.43
40	5400	480	4920	0.67
50	4500	600	3900	1.00
60	3600	720	2880	1.50
70	2700	840	1860	2.30
80	1800	960	840	4.00
90	900	1080	−180	9.00
100	0	1200	−1200	

[a]Assuming 500°F stack temperature.

Reprinted by permission of Johnson, A. J., and G. H. Auth, 1951. Table 5-9 in *Fuels and Combustion Handbook*, McGraw-Hill, New York.

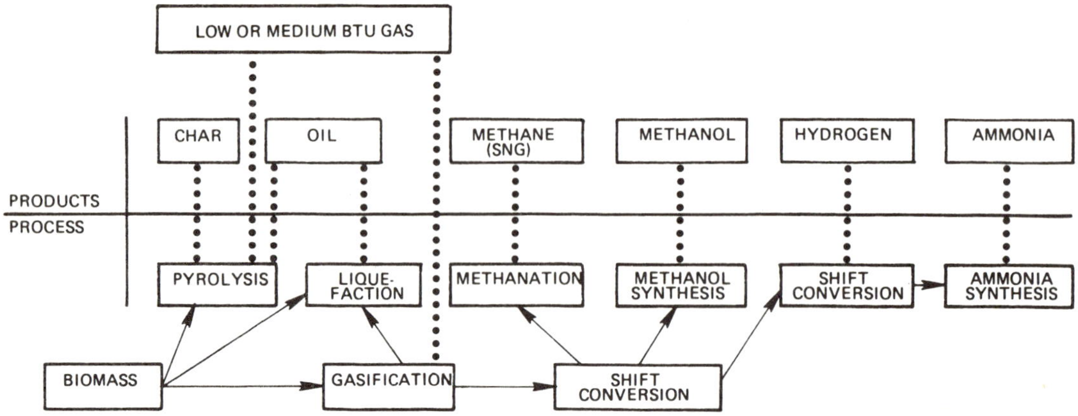

Fig. 4.2. Products generated by thermochemical conversion processes.

gies are commercially available (Kohan and Barkhorder 1979).

Table 4.19 identifies five specific processes as examples of the three categories of thermochemical conversion and lists the major features of each process.

4.2.2.1 Pyrolysis. Pyrolysis is the thermal decomposition of biomass materials. Strictly speaking, this decomposition must occur in the absence of air; however, some pyrolytic processes use air in the reactor vessel. If a reactor uses air during operation and maintains a stage, or zone, in which an anaerobic thermal decomposition occurs, the reactor can remain classified as pyrolytic. Pyrolyzers operate at approximately atmospheric pressures and at temperatures adequate to volatilize the biomass to liquids and gases. Liquids are the main product when temperatures are controlled around 205°C, and primarily gases are produced at temperatures near 650°C (Garrett 1977*a* and Kohan and Barkhorder 1979).

As shown in Fig. 4.2, pyrolytic processes can produce char, oil, and low- or medium-Btu gas. The two pyrolytic conversion processes identified, one developed by the Garrett Energy Research and Engineering Company and the other by the Tennessee Valley Authority (TVA), use different feedstocks, reactor types, and operating conditions and produce different products. Details are described in Sects. 4.5.2.2 and 4.5.2.3.

4.2.2.2 Liquefaction. Liquefaction is a process in which carbonaceous material loses oxygen through a reaction with carbon monoxide. On loss of oxygen and possible addition of hydrogen from either water or pure hydrogen, the material is converted into an oil. As shown schematically in Fig. 4.2, liquefaction may be achieved by direct processing of biomass or it may proceed through a preliminary gasification step. Liquefaction processes are designed to run at high pressures and temperatures and, in some cases, with alkali catalysts. A temperature range of 370 to 400°C is usual for operating an uncatalyzed liquefaction process; however, reasonable, or rapid, reaction rates are achieved below 315 to 345°C by using alkali catalysts such as sodium carbonate (Kohan and Barkhorder 1979).

4.2.2.3 Gasification. During gasification, biomass is converted to a mixture of gases containing carbon monoxide and hydrogen as the primary combustibles. Gasification is based on incomplete oxidation of the feedstock materials, usually producing either low- or medium-Btu gas. Low-Btu gas is produced if air is used in the gasifier, and medium-Btu gas is obtained if pure oxygen is used. Although not shown in Fig. 4.2, the product gas stream usually must be purified before it can be stored, transported, or used as a feedstock for other processes. Gasifiers often operate at temperatures and pressures greater than 538°C and 2 MPa (300 psia).

Following purification, the product gas from a gasifier can be added to pipeline gas; however, current regulations prohibit this practice. The product gas can be used as a chemical feedstock for other useful fuels or chemicals by entering a refining scheme, the first step of which is a shift conversion (Fig.

Table 4.19. Thermochemical conversion processes

Process developer	Process type	Feedstock	Operating conditions	Major equipment	Products
Garrett Energy Research & Engineering Co., Inc. (GERE)	Pyrolysis	Manure, sawdust, or cotton gin trash	$T = \sim 1200°F$ P = atmospheric	Multiple-hearth furnace	Medium-Btu gas
TVA, Maryville	Pyrolysis	Forest and agricultural residue	$T = \sim 400°F$ P = atmospheric	Mobile vertical flow reactor	Low-Btu gas, oil, and char
Bechtel Corp. (PERC process)	Liquefaction	Wood	$T = \sim 650°F$ $P = \sim 2000$ psia	High-pressure reactor	Oil
Wright-Malta	Gasification (catalytic)	Wood, peat, cornstalks	$T = \sim 1100°F$ $P = \sim 500$ psi	Pressurized indirectly heated rotary kiln	Hydrogen, methane, and light hydrocarbons
Texas Technological University	Gasification	Manure, wood, corn stover, mesquite cotton gin trash, wheat straw	$T = \sim 650°C$ P = atmospheric	Variable-velocity fluidized bed	Low- and medium-Btu gas

Sources: Kohan, S. M., and P. M. Barkhorder, 1979. *Mission Analysis for the Federal Fuels from Biomass Program*, vol. IV, prepared by S.R.I. International for the Division of Distributed Solar Technology Fuels from Biomass Systems Branch, U.S. Department of Energy, Washington, D. C.; Garrett, D. E., 1977. *Conversion of Biomass Materials into Gaseous Products, Phase 1*, SAN/1241-77/1, Garrett Energy Research and Engineering, Inc., Claremont, Calif.

4.2). In shifting, water is allowed to react with carbon monoxide to form hydrogen and carbon dioxide. This reaction is controlled so that the product gas from this step has hydrogen to carbon monoxide ratios of 2 to 1, 3 to 1, or greater. A 2-to-1 mixture is used for methanol production, and a 3-to-1 mixture is used for methane production. A nearly complete shift to hydrogen is required for ammonia production and is accomplished in a series of shift converters, with carbon dioxide removal occurring between the converters as illustrated in Fig. 4.2.

Two process developers who are using gasification are listed in Table 4.19. The Wright Malta gasification process uses a catalyst in the gasifier that converts carbon monoxide directly to methane. The process is at the bench scale of development and operates at a high pressure. Texas Technological University is operating a pilot-scale unit, as described in Sect. 4.5.2.1.

4.2.3 Biochemical Conversion

Figure 4.3 is a flow diagram of the two principal biochemical conversion processes, anaerobic digestion and fermentation, and their associated products. Both processes are anaerobic and both involve fermentation. The distinction lies in the microorganisms, their digestion activities, and the resulting products. Anaerobic digestion is probably most accurately described by the designation *methane fermentation,* whereas what is usually called fermentation is actually alcohol fermentation. Hydrolysis is included as a process in this section because it is used almost exclusively in association with fermentation; hydrolysis is described in greater detail in Sect. 4.5.4.1.

4.2.3.1 Hydrolysis. Hydrolysis refers to the breakdown of complex organic material such as cellulose to soluble sugars. This breakdown occurs as a separate processing step in the overall conversion scheme and is commonly used in producing sugars for fermentation. Hydrolysis is accomplished by reacting biomass with water in the presence of acid or enzymes.

Lignin inhibits both the acid or enzyme hydrolysis reaction and remains insoluble after the reaction. Sodium hydroxide has been shown to be an effective delignifier, but it causes the lignin to hydrolyze to alcohols and acids when operating at elevated temperatures (Jones and Fong 1978).

4.2.3.2 Anaerobic digestion. Anaerobic digestion is a multistage process in which biomass is converted to methane. Potential feedstocks for anaerobic digestion are manure, wheat straw, and ocean kelp. Of these three, manure from cattle feedlots appears to be the best source of material for digestion because it is high in moisture content and in biodegradable material and seldom requires additional nutrients.

The first stage of digestion is an enzymatic hydrolysis of the biomass to sugars by enzymes that are produced by bacteria normally present under anaerobic conditions. The sugars are converted to organic acids by bacteria in the second stage of the digestion process. The final stage occurs when another

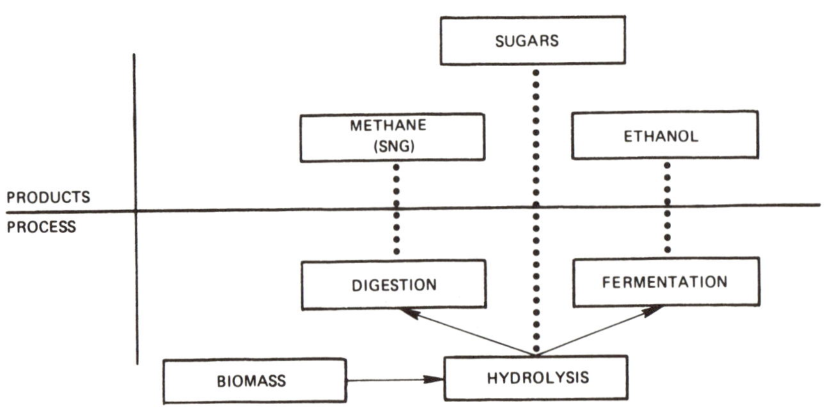

Fig. 4.3. Products generated by biochemical conversion processes.

set of bacteria converts the acids to methane in the presence of carbon dioxide and hydrogen. Figure 4.4 illustrates this process and lists the various process by-products. The steps shown in Fig. 4.4 occur simultaneously, which allows for the design of continuous flow operations.

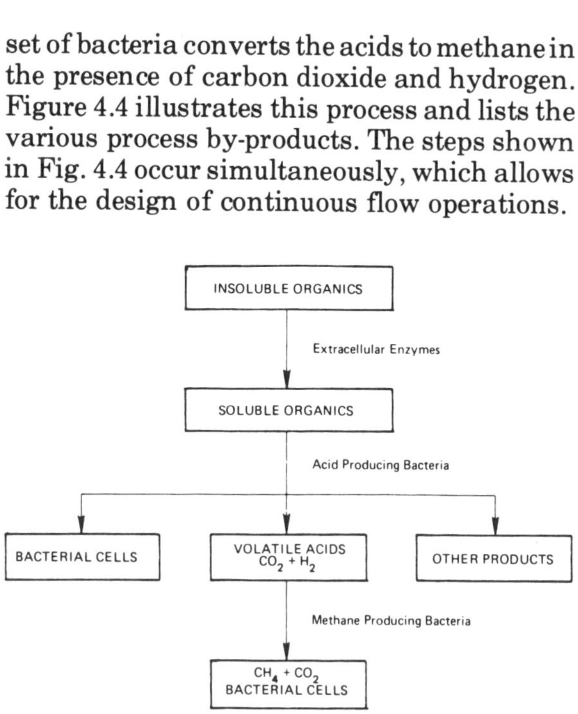

Fig. 4.4. Anaerobic digestion process. Source: Jones, J. L., and W. S. Fong, 1978. *Mission Analysis for the Federal Fuels from Biomass Program*, vol. V, prepared for the Division of Solar Energy Fuels for Biomass Systems Branch, U.S. Department of Energy, by S.R.I. International, Washington, D.C.

Properly designed digestion processes generally maintain the process pH between 6 and 8 and the temperature within one of two ranges: mesophilic (20 to 40°C) or thermophilic (45 to 65°C). To maintain the temperature, most digesters are insulated and heated by using the produced methane as the fuel source. Ammonia or antibiotics can inhibit the digestion process, and nutrients may be required for certain feedstock materials. Details of full-scale anaerobic digestion are given in Sect. 4.5.3.

4.2.3.3 Fermentation. The production of ethanol via fermentation has been practiced for centuries. During fermentation, yeast converts sugar to ethanol in concentrations up to 15 vol %. Distillation is required if pure or higher concentrations of ethanol are desired.

Currently, natural sugars and starches make up the feedstock for most of the fermentation ethanol, which is produced almost exclusively for human consumption. Industrial or analytical ethanol is produced from ethylene, a petroleum-derived feedstock. Because sugars and starches are relatively expensive and more valuable as consumable goods, cellulosic materials are being investigated as alternative feedstocks.

In general, the pH of a fermentation process ranges between 4 and 6 the temperature between 18 and 21°C. The sugar feedstock solution has a concentration range of 10 to 20 wt % (Jones and Fong 1978). Industrial-scale fermentation is described in Sect. 4.5.4.1.

4.3 End Products

Although this book does not include environmental effects of end products or end uses, the products of conversion are briefly described as they relate to the processes and to their potential impact on the environment as they are produced or stored on the production site.

Conversion of biomass materials can give rise to a number of fuels and chemicals. The most important of these in relation to near-term accessibility are synthesis gas, methane, and ethanol. Synthesis gas is the product of biomass gasification, methane is produced by either bacterial anaerobic digestion or thermochemical degradation, and ethanol is the product of fermentation of biomass. The products can be used directly as fuels or can be converted into other forms—synthesis gas into methane or methanol, methane into ammonia for fertilizer, and ethanol into ethylene (a chemical feedstock).

4.3.1 Synthesis Gas

Synthesis gas, which is usually a stoichiometric mixture of hydrogen and carbon monoxide or carbon dioxide, gets its name from its use in the synthesis of methanol, although it also provides the basic components for the synthesis of methane, ammonia, and the various chemicals derived from these. Synthesis gas can be obtained by processing the crude gas obtained from the gasification of any carbonaceous biomass material such as wood, wood waste, crop residues, or animal manures. Depending on feedstock and gasification conditions, the crude gas will contain, in addition to hydrogen and carbon monoxide, varying amounts of carbon dioxide, methane, nitrogen, oxygen, or other hydrocarbons. Examples of gas compositions from biomass gasification are listed in Table 4.20. Clearly, a broad range of compositions is possible, broader for hydrogen than for CO.

The composition of gas used to produce methane (pipeline gas) or methanol must be "shifted" to a 3-to-1 or 2-to-1 ratio of hydrogen to carbon monoxide. This shift is accomplished by passing part of the gas stream through a carbon monoxide converter, where steam is added to react with the CO to produce CO_2 and H_2 by employing the well-known water-gas shift reaction. The shifted gas is remixed with the original, and some or all of the CO_2 is removed.

4.3.2 Methane

Methane is a colorless, odorless gas with less than one-half the density of air. It is the major component of natural gas, constituting 75 to 95% of the mixture. Because of the high hydrogen-to-carbon ratio in methane (four hydrogen atoms for each atom of carbon), it is the most environmentally desirable carbon-containing combustion fuel. [Hydrogen (H_2) is the most desirable noncarbon combustion fuel because it produces only heat and water vapor as combustion products: $H_2 + ½O_2 = H_2O$ + heat.] Compared with the other saturated light hydrocarbons, methane has the highest Btu value per unit of mass and the lowest CO_2 emission rate per unit heat input—88% the rate of its nearest competitor, ethane, which produces 30 m^3 per 10^9 J input (1.12 ft^3 of CO_2 per 1000 Btu) (Johnson and Auth 1951). In addition to its value as a fuel, methane is heavily used as an industrial feedstock in processes such as ammonia (fertilizer) production, synthesis of organic chemicals, and carbon black production. Carbon black is used in large quantity as a pigment in printer's ink and as the filler and reinforcing agent in making rubber. (See Sect. 4.1.2 for reference to PAHs in carbon black.)

As a major component of natural gas, methane occurs abundantly in nature, often trapped in subterranean cavities in association with coal deposits. It also occurs naturally as a normal product of microbial degradation of swampy vegetation. Called "marsh gas," it is continually vented to the atmosphere. Pure methane is obtained from natural gas by liquefying and separating the heavier compounds in the gas as liquefied natural gas (LNG).

Methane burns readily in air with a pale, slightly luminous, very hot flame. Although methane gas is stable, it is the primary cause of mine disasters, forming explosive mixtures with air at concentrations as low as 5 to 14%.

Table 4.20. Composition of crude gas from various processes (vol %)

Components	Texas Tech process (manure)[a]	Texas Tech process (oak sawdust)[b]	Moore-Canada (wood waste)[c]	Molten salt process (rubber)[d]	Molten salt process (sugar)[d]	Steam-gasification (wood, sawdust)[e]
Hydrogen	28.2	2.8	18.3	16.0	18.9	40.9
Carbon monoxide	16.4	29.6	22.8	18.4	26.9	15.1
Carbon dioxide	15.4	29.3	9.2	4.0	13.5	28.5
Methane	9.2	8.8	2.5	2.4	4.3	14.0
C_2-hydrocarbons	7.6[f]	3.4[g]	0.9	1.1	1.9	1.5
Nitrogen	23.2	23.6	45.8	Not reported	34.3	

[a]Hoffman, W. S., J. E. Halligan, and R. C. Peterson, 1978. *Conversion of Cattle Feedlot Manure to Ethylene and Ammonia Synthesis Gas*, EPA-600/2-78-026, Environmental Protection Agency, Washington, D.C.

[b]Beck, S. R., et al., 1979. "Application of SGFM Technology to Other Feedstocks, Phase II," presented at 6th Biomass Thermoconversion Contractors Meeting, Tucson, Ariz., Jan. 16—17, 1979.

[c]Hokanson, A. E., and R. M. Powell, 1977. *Methanol from Wood Waste: A Technical and Economic Study*, General Tech. Rep. FPL12, Forest Products Laboratory, Forest Service, U.S. Department of Agriculture, Washington, D.C.

[d]Yosim, S. J., and K. M. Bardlay, 1977. "Production of Low-Btu Gas from Wastes, Using Molten Salts," in *Fuels from Waste*, L. L. Anderson and D. L. Tillman, eds., Academic Press, New York.

[e]Coffman, J. A., 1979. "Catalyzed Steam Gasification of Biomass, Phase II," presented at 6th Biomass Thermoconversion Contractors Meeting, Tucson, Ariz., Jan. 16—17, 1979.

[f]Ethylene, 4.9%; ethane, 2.7%.

[g]+2.4% of C_3 + hydrocarbons.

Although nontoxic, methane can produce asphyxiation when present in concentrations high enough to cause oxygen deprivation. Again, this observation is based on incidents in underground, unventilated coal mines. Methane is a normal, though minor, component of the atmosphere, with concentrations that vary strongly with location, and it is the only atmospheric hydrocarbon not implicated in the production of photochemical smog.

4.3.3 Methanol

Methanol, also known as methyl alcohol, wood alcohol, carbinol, and wood spirit, is a clear, colorless, mildly odorous liquid that boils at 65°C (149°F). When distilled from wood, methyl alcohol has the characteristically pungent odor of its contaminants, acetone, acetic acid, and allyl alcohol.

Methanol is currently produced commercially from synthesis gas by passing hydrogen and carbon monoxide over a catalyst (chromium or zinc and copper oxides) at elevated temperatures and pressures (Hagen 1976). Methanol can also be synthesized from methane by direct oxidation in the presence of a selective catalyst. Methanol has important motor fuel properties, compared with isooctane and gasoline (Table 4.21). Unlike gasoline, methanol burns with a very hot nonluminous, and therefore invisible, flame; accordingly, accidental ignition of the fuel may not be detected immediately.

McKetta and Cunningham (1978) list four advantages to using pure methanol as a motor fuel: (1) high power output, (2) high octane rating, (3) cooling effect, and (4) miscibility with power-boosting nitroparaffins. The high octane number of methanol makes it possible to use compression ratios as high as 15 to 1, but the high latent heat of vaporization and the low air-fuel ratio are problems in starting and acceleration. Compared with gasoline, meth-

Table 4.21. Physical properties of fuel

Property	Methanol	Isooctane	Gasoline
Heat of combustion			
Btu/lb at 77°F			
Liquid fuel—gaseous H_2O	8,593	19,065	18,900
Gaseous fuel—gaseous H_2O	9,100	19,195	19,050
Btu/gal at 77°F			
Liquid fuel—gaseous H_2O	56,700	110,000	115,300
Gaseous fuel—gaseous H_2O	60,100	110,800	116,200
Heat of vaporization, Btu/lb at 77°F	503	132	150
Density at 68°F, g/ml	0.791	0.692	
Density at 68°F, lb/gal	6.60	5.77	6.1
Boiling point, °F	148.1	210.6	80—437
Vapor pressure at 100°F, lb/in.^{2}	4.63	1.71	7—15
Octane number			
Research	106	100	91—100
Motor	92	100	82—92
Flame propagation speed, ft/s	2.5	1.9	
Flammability limits, vol %			
Lower	6.7	1.1	1.4
Higher	36.0	6.0	7.6
Auto ignition temperature, °F	867	784	430—500
Stoichiometric mixture, lb air/lb fuel	6.44	15.1	14.2—14.8
Energy content, Btu/ft^3 of stoichiometric mixture, gaseous fuel—gaseous H_2O	91.1	92.4	~93

Reprinted by permission of McKetta, J. J., and W. A. Cunningham, 1978. *Encyclopedia of Chemical Processing and Design*, vol 7, Marcel Dekker, Inc., New York.

anol requires eight times as much heat per volume of air-fuel mixture to obtain complete vaporization. In addition to using it in its pure form as a combustion fuel, methanol can be mixed with gasoline in a 10% solution to make gasohol, a liquid motor fuel with properties that do not differ strongly from those of pure gasoline. Methanol is also an important source for synthesis of formaldehyde, a feedstock for synthetic resins.

Handling and storage of methanol must include protective precautions because of its flammability. Steel or reinforced concrete storage tanks are normally located some distance from the heart of an industrial plant area. The tanks are surrounded by a dike and protected by a foam-type fire extinguishing system. Flammability limits are 6.7 to 37% by volume. Inhalation of methanol vapors can cause symptoms of poisoning at concentrations of 1000 ppm and profound narcosis at 50,000 ppm. The maximum allowable limit for exposure for an 8-h working day is 200 ppm.

4.3.4 Ethanol

Ethanol, or ethyl alcohol, is a clear, colorless, flammable liquid with a mild odor and a pungent taste (unless suitably diluted). Industrially, ethanol is produced either petrochemically from ethylene or by fermentation of sugar or starch. Although other synthesis routes are available, few have proved commercially successful. One of these—the Fischer-Tropsch condensation from synthesis gas, carbon monoxide, and hydrogen—is technically, but not yet economically, feasible.

As a motor fuel, pure ethanol results in engine performance and fuel economy that lie between those of gasoline and pure methanol. In one study comparing exhaust emissions, the use of ethanol produced lower carbon monoxide emissions than did methanol or gasoline, higher unburned hydrocarbon emissions, and nitrogen oxide emissions higher than those for gasoline but lower than those for methanol (Ecklund 1978). For a 20% blend of alcohol in gasoline, the performance and emission characteristics of an ethanol-gasoline blend were found to fall between that of a methanol-gasoline blend and pure gasoline. It was concluded that practical blends of alcohol with gasoline (10 to 15 vol %) would have little effect on steady-state automobile engine performance and currently regulated emission characteristics (Ecklund 1978).

4.4 Effluents and Emissions

The effluents from biomass conversion are largely undefined. Because much of the new technology is still in either a developmental or a pilot stage and because the environmental impacts were not expected to be as severe as those from fossil fuel technologies, less attention has been given to waste streams and emissions than to process development. However, as newer processes approach commercialization, the development of more precise waste and effluent information will become increasingly important.

In the interim, existing information, along with some gleaned from kindred technologies, may be useful for uncovering areas of possible environmental significance. Additionally, knowledge about the components of feedstock materials (Sect. 4.1) is valuable for detecting possible sources of environmental emissions. Figures 4.5 and 4.6 give a schematic overview of the types of wastes to be expected from conversion processes.

4.4.1 Combustion

The major emissions from almost any means of combustion of biomass materials are air pollutants—notably, particulates and carbon monoxide—with lesser amounts of nitrogen oxides, sulfur oxides, and hydrocarbons. The amounts of these depend on a number of variables, such as the composition of the fuel, the furnace or stove design, and the combustion conditions (e.g., feed rate and amount of excess air).

Emissions are strongly dependent on operating conditions. For example, proper drying conditions must be supplied to reduce the high moisture content of the fuel prior to combustion; otherwise, the combustion temperature will be lowered, causing incomplete combustion with high emissions of particulates and carbon monoxide (Sect. 4.4.2.1). Additionally unless sufficient secondary air is supplied above the fuel bed to completely burn the volatiles, a high hydrocarbon emission will also result. Wood is estimated to contain 80%

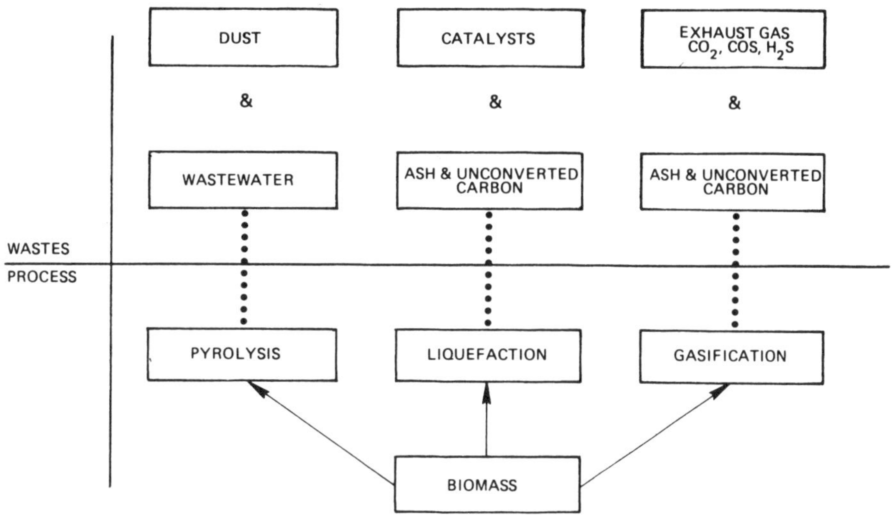

Fig. 4.5. Wastes generated by thermochemical conversion processes.

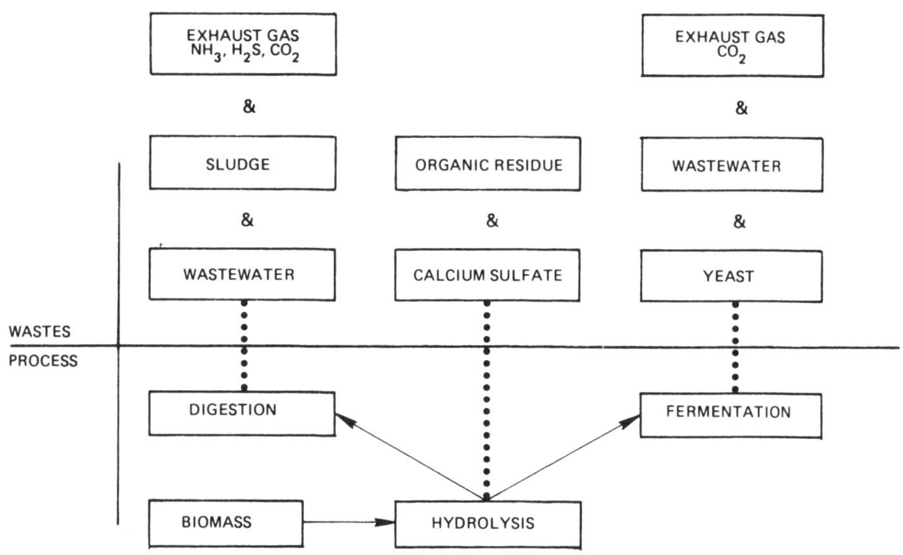

Fig. 4.6. Wastes generated by biochemical conversion processes.

volatiles by dry weight. For this reason, home fireplaces that have no provision for burning volatiles are not only inefficient heating devices (less than 15% efficiency) but can be serious polluters (Sect. 4.5.1.3). Likewise, open burning, in which low temperatures and poor air circulation are usual, leads to high particulate, carbon monoxide, and hydrocarbon emissions (Table 4.22).

4.4.1.1 Nitrogen oxides. Nitrogen oxides (NO and NO_2, often referred to as NO_x) are often emitted from combustion processes. Although these oxides can definitely be produced from the nitrogen present in the fuel (Martin and Berkau 1972), nitrogen itself is not flammable. Rather, most NO_x emitted from combustion processes is formed from the atmospheric nitrogen (N_2) in the air feed, which is converted to NO_x in the high-temperature zones within the combustion region. Nitrogen oxides formed in this way are termed *thermally produced.* The formation

Table 4.22. Air emission factors from combustion of biomass (lb/ton)

Source	Particulates	Carbon monoxide	Hydrocarbons	Sulfur oxides	Nitrogen oxides
Wood/bark in boilers, 50%	30[a]	2–60	2–70	1.5	10
Wood in residential fireplaces	20[b]	120	5	0	1
Small wood stoves, oak and pine	3.4–20.0				
Waste open burning, agricultural and wood refuse					
Unspecified	21	117	23		
Corn	14	108	16		
Rice (dry)	9	83	10		
Rice (wet)	29	161	21		
Sorghum	18	77	9		
Sugarcane	7	71	10		
Oats	21–44	136	18–33		
Wheat	13–22	108–128	11–17		
Forest residues, unspecified	17	140	24		4
Bagasse burning	22				
Bagasse boilers	16				

[a]Bark only, 50 lb/ton.
[b]Contains as much as 70% condensate.
Source: Compiled from data in U.S. Environmental Protection Agency, 1973. *Compilation of Air Pollutant Emission Factors*, AP-42, 2d ed. (and supplements), Research Triangle Park, N.C.

mechanisms for this conversion are complex (Lange 1972) and not yet fully understood. Control of the operating conditions and design parameters of the combustion system is the most effective means of regulating thermally produced NO_x emissions. Two mechanisms appear capable of reducing the conversion of atmospheric N_2 to NO_x: (1) reduction of the peak flame temperature by flue gas recirculation or (2) control of the swirl momentum of the intake air. However, conversion of fuel-bound nitrogen is much less sensitive to temperature and recirculation of air. The NO_x formation from fuel nitrogen depends on competing reactions of nitrogen-containing intermediates to form NO with oxygen or N_2 by recombination. The percentage of fuel-bound nitrogen converted to NO has been reported as between 20 and 70% (Martin and Berkau 1972). Thus the intrinsic nitrogen content of biomass combustion fuels is important in determining NO_x emissions (Sects. 4.1.2.2 and 6.1.1).

In a recent study (Nelson et al. 1979) in which wheat straw was cofired in a 50/50 mixture with coal in an experimental unit simulating a pulverized coal utility boiler and in which standard NO_x control techniques (e.g., whirl) were employed, it was found that the addition of wheat straw to coal caused the nitrogen oxide emissions to increase fourfold over firing with coal alone. This occurred in spite of the fact that the fuel-bound nitrogen in the straw was far less than that in the coal—0.55% in the wheat straw compared to 0.84% in the coal (Nelson et al. 1979). It was not clear whether the increased emissions could be accounted for by any one or a combination of the following possible explanations.

1. Nitrogen in wheat straw is held in weakly bound functional groups (compared to coal) and thus would be released more readily in the flame region where it could be oxidized to NO.

2. Wheat straw has a much higher volatile-matter content than coal (71% in straw vs 26% in coal) and nitrogen oxide emissions are known to increase with increased volatile content of a fuel.

3. Because wheat straw, like most cellulosic biomass, has a low-density, porous structure in which air is entrained, it burns rapidly and intensely and is capable of producing localized hot spots, high temperature regions, which are known to promote the conversion of atmospheric N_2 to NO.

Clearly, there is a need for systematic, detailed studies of NO_x emissions from biomass combustion in industrial boilers.

4.4.1.2 Carbon monoxide. Carbon monoxide is readily formed in the combustion of almost any carbonaceous fuel in a limited supply of air or oxygen according to the reaction

$$C + \frac{1}{2}O_2 = CO . \quad (4.1)$$

If excess air or oxygen is available at the combustion site, the reaction will be driven to complete combustion,

$$C + O_2 = CO_2 . \quad (4.2)$$

However, feeding excess air through a solid-fuel fixed bed will not prevent the copious production of carbon monoxide as in burning wood, wood waste, or packed biomass in a wood-burning stove or fireplace; in fact, it will increase it. This occurs because the CO_2 formed as indicated in Eq. (4.2) is quickly reduced by contact with the glowing coals in the bed (Kreisinger 1939) according to the reaction

$$C + CO_2 = 2CO. \quad (4.3)$$

Equation (4.3), which represents an equilibrium between the two gas-phase species, CO and CO_2, and the solid-phase carbon (C), is slightly endothermic as written, absorbing about 170 kJ/mole (160 Btu/mole). The rate of the reaction is strongly temperature dependent, proceeding raipdly and efficiently to completion above 1000°C (1830°F) but proceeding more slowly below 800°C (1470°F). The temperature range 800 to 900°C (1470 to 1650°F) corresponds to a bed of embers glowing bright cherry red to orange (Table 4.23). The reaction is reversible so that as the products of combustion cool and come into contact with a solid surface, some of the carbon monoxide can disproportionate, depositing carbon on the internal surface of stovepipes and chimneys. Thus, if the carbon monoxide is not ignited and burned above the bed, it will be emitted either as the toxic gas or, if converted, as particulate carbon (soot).

Table 4.23. Approximate color of glowing hot, solid objects

Appearance	Temperature	
	°F	°C
No emission detectable	<885	<475
Dark red	885–1200	475–650
Dark red to cherry red	1200–1380	650–750
Cherry red to bright cherry red	1380–1500	750–815
Bright cherry red to orange	1500–1650	815–900
Orange to yellow	1650–2000	900–1090
Yellow to light yellow	2000–2400	1090–1315
Light yellow to white	2400–2800	1315–1540
Brighter white	>2800	>1540

In analyzing the situation, Kreisinger (1939) notes: "The important deduction is that all of the combustion of the volatile matter, which constitutes over 80 per cent of the combustible in the wood-waste fuel and about a half of the combustion of the fixed carbon, takes place above the fuel bed. Therefore, about 90 per cent of the air required for combustion should be supplied over the fuel bed and only about 10 per cent through the fuel bed." Unfortunately, air alone above the bed is usually insufficient for burning the CO unless it is heated to 593°C, the ignition temperature of carbon monoxide.

4.4.1.3 Hydrocarbons. Combustion is an important source of hydrocarbon emissions (Table 4.22). Many of the components of hydrocarbon emissions have been identified but not completely quantified (Sect. 4.4.2.1). Some polycyclic hydrocarbons have been reported in wood smoke (Table 4.24), although no definite relationship to a particular fraction of wood is known (see also Sect. 4.1.2). Aromatic hydrocarbons have been reported in wood fly ash collected during operation of a wood-fired utility boiler (Sect. 4.5.1.1). Studies to clearly identify and to quantify hydrocarbon emissions from biomass combustion processes are badly needed.

4.4.2 Thermochemical Conversion

Because the thermochemical conversion of biomass includes a variety of technologies (Sects. 4.2.2 and 4.5.2) and feedstock materials (Sect. 4.1), a wide range of emissions is to be expected. Little direct information exists that describes the effluents from thermochemical conversion. Some possible effluents and emissions are described in Sect. 4.5.2 in connection with the full-scale facility descriptions. Additionally, specific effluents that may be associated with a particular feedstock are described in Sect. 4.1. This section discusses studies of known effluents from existing technologies that use biomass (though not necessarily for conversion to energy). In

Table 4.24. Some compounds reported to be present in hardwood smoke

Acids	Phenols
Formic	Cresols
Acetic	Creosol
Propionic	Guaiacol
Butyric	Guaiacol derivatives
Tricarballylic?	4-Ethyl
Ketoglutaric?	4-Propyl
	6-Methyl
Alcohols	6-Ethyl
Methanol	6-Propyl
Ethanol	Pyrogallol ethers
	1-0-Methyl
Carbonyls	1,3-Dimethyl
Formaldehyde	1,3-Dimethyl pyrogallol derivatives
Acetaldehyde	5-Methyl
Acetone	5-Ethyl
Diacetyl	5-Propyl
Furfural	1-0-Methyl-5-methyl pyrogallol
Methyl furfural	Veratrole
	Xylenols
Hydrocarbons	
3,4-Benzpyrene	Ammonia
1,2,5,6-Dibenzanthracene	Carbon dioxide
20-Methylcholanthrene	Resins
	Water
	Waxes

Source: Jahnsen, V. J., 1961. *The Chemical Composition of Hardwood Smoke*, Ph.D. thesis, Purdue University, Lafayette, Ind.

particular, the effluents in wood smoke are described, because smoke can be associated with the thermal processing of almost any cellulosic material.

4.4.2.1 Smoke. Smoke is commonly emitted during biomass conversion, but little information on the characteristics of this smoke is available. However, wood smoke (especially hardwood smoke) has been studied in some detail in the laboratory (King 1975; Einhorn et al. 1973; Doerr et al. 1966; Jahnsen 1961). Smoke from other biomass materials may not be too different from that of wood, but this fact will need to be confirmed. Two types of studies were encountered, those that focus on hydrocarbons in wood smoke and those that examine carbon monoxide and smoke. Health effects are implicated in each.

Wood smoke is a complex mixture of solid particles, condensed liquid particulates, and volatiles and is produced most abundantly by combustion of moist wood in a restricted flow of air (Jahnsen 1961). Figure 4.7 shows the amount of wood smoke condensates produced at different wood moisture contents for hickory sawdust. Clearly, the volume of airborne condensate increases rapidly and linearly with increasing levels of moisture. Of the 22 hydrocarbon compounds identified in the condensate, which contained acids, alcohols, carbonyls, and phenols, formic and acetic acids constituted 57%, ethanol 10.3%, and 2,6-dimethoxy pyrogallol 4.2%. The remaining fraction contained furfural, anisole, the three cresols, veratiole, xylenols, hydroquinone, thymol, vanillin, guaiacol, pyrocatechol, pyrogallol and phloroglucinol. In order of decreasing concentration, the carbonyls identified were acetone, acetaldehyde, formaldehyde, furfural, and several diacetyls.

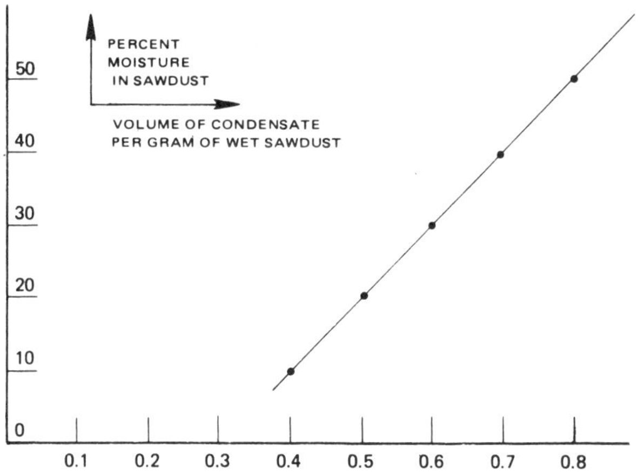

Fig. 4.7. Relationship of sawdust moisture content to the volume of smoke condensates. Source: Jahnsen, V. J., 1961. *The Chemical Composition of Hardwood Smoke*, Ph.D. thesis, Purdue University, Lafayette, Ind.

In another study of smoke from hickory sawdust (Doerr et al. 1966), among the major components of whole smoke were the gases, carbon monoxide, carbon dioxide, and methane, and the condensables furan and propionaldehyde.

The occupational health standards for concentrations in air of substances produced by the thermal decomposition of wood (e.g., hickory sawdust) are as follows (Ramos et al. 1972):

Acetaldehyde	200 ppm
Acetone	1000 ppm
Isopropyl alcohol	400 ppm
Methyl formate	100 ppm
Methyl alcohol	200 ppm
Total dust	15 mg/m^3 of air

The relationship between wood smoke and the emission of carbon monoxide is important. A common myth is that, when the concentration of smoke particles is low, so is the concentration of combustible gases (King 1975). However, in a controlled study that compared the formation of smoke and carbon monoxide from burning red oak with their formation from burning other flammable organic materials, red oak liberated large quantities of carbon monoxide under nonflaming burning conditions. The most rapid rate of carbon monoxide generation (80 ppm/min for a 4.3-g sample heated at 2.5 W/cm^2) occurred with the onset of glowing combustion (see Sect. 4.4.1.2). Unlike the other flammables, which produced far more smoke than carbon monoxide under the test conditions, red oak had the highest carbon monoxide emission rate and the lowest smoke production (King 1975). The smoke was found to be a mixture of solid particles and elliptical liquid residue, with particle size ranging from 0.2 to 0.7 μm.

4.4.2.2 Wood-dryer emissions. The stack gas condensate from a Douglas-fir veneer dryer was investigated to characterize the organic emissions and to evaluate their toxicity (Fraser and Swan 1972). The condensate was about one-third acids (resin acids and fatty acids) and two-thirds neutrals, which were poorly characterized but reported as mainly composed of lipids with more than 15 carbon atoms. Samples from a smoking stack contained more resin acids than those from a clean stack, suggesting that a smoking-stack gas contains more condensate hydrocarbons, whereas a clean-stack gas may contain more volatile hydrocarbons. The possible environmental effects of scrubbing the stack gases to remove the condensable hydrocarbons is discussed in Sect. 6.2.

4.4.2.3 Wastewater effluents from the wood-products industry. In the absence of direct data on wastewater effluents from biomass conversion technologies, guidelines for control of wastewater effluents from the wood-products processing industry (Thomson 1976) are useful for examining possible effluents from the preconversion portion of biomass conversion technologies. For example, wet storage of logs produces floating solids in the holding stream, in addition to tannins and other wood extractables. Log washing and debarking results in suspended solids and biological oxygen demand (BOD). In addition, water or steam in contact with wood or wood products leads to an aqueous discharge high in BOD and phenols (Thomson 1976). Inasmuch as wastewater effluents from the wood products industry are treatable using conventional water treatment methods, it is expected that wastewater from thermochemical processing of biomass will also be treatable.

4.4.3 Anaerobic Digestion

The wastes from anaerobic digestion are often thought to be manageable if not actually desirable. This is because the biomass feedstock is generally animal manure, which is environmentally unacceptable, whereas the waste sludge is odor free and considered marketable as refeed material, crop fertilizer, and soil conditioner. The use of wastes from anaerobic digestion is discussed in Sects. 3.3 and 5.3. However, the question of amplification of trace element concentrations as a result of refeed is addressed in Sects. 4.1.7 and 6.3.

4.4.3.1 Air emissions. Digester gas almost always contains traces of hydrogen sulfide. Hydrogen sulfide is very corrosive and, if allowed to remain in the fuel gas, may corrode storage and piping facilities, in addition to corroding the contacting part of an internal combustion engine (Kuo and Jones 1978). In addition, the release of hydrogen sulfide has a potential for causing nuisance odors.

The pollution potential of an anaerobic digestion system has been estimated by Kuo and Jones (1978). Air emissions from the entire process are listed in Table 4.25. The sulfur and nitrogen emissions arise from the combustion of H_2S and NH_3 in the product gas as it is burned in a boiler for digestion-system heating. The NH_3 content of the gas depends on the pH of the digestion slurry, reaching about 2% at a pH of 7.5. Both H_2S and NH_3 are expected to be emitted from stored manure, sludge, and gas (from tank leakage).

4.4.3.2 Wastewater emissions. Digester effluent may be separated into a liquid and a sludge. Both will require storage and handling facilities because they are high in BOD and, although they have a less offensive odor than manure, are considered pollutants (Thornton 1978).

Potential water emissions include manure-storage runoff water, sludge-storage runoff water, and any process water not recycled. Table 4.26 gives the relative amounts of wastewater per pound of dry-solids feed as calculated by Kuo and Jones (1978).

4.4.4 Fermentation

The emissions and effluents associated with fermentation are described in Sect. 4.5.4.

Table 4.26. Relative amounts of wastewater per pound of dry-solids feed

Feedstock[a]	Output (lb/lb dry solid feed)		
	Dry residue	H_2O	Wastewater
Cattle manure, fresh	0.614	1.842	2.158
Cattle manure, 5-month collection cycle	0.837	2.511	0
Cattle manure, fresh plus wheat residue	0.706	2.118	0.582

[a]Each entry corresponds to a different case.

Reprinted by permission of Kuo, T. M. C., and J. L. Jones, 1978. "Environmental and Energy Output Analysis for the Bioconversion of Agricultural Residues to Methane," pp. 143—76 in *Proc. of the IGT Conf. on Energy from Biomass and Wastes, Aug. 14—18, 1978, Washington, D.C.*, Institute of Gas Technology, Chicago, Ill.

Table 4.25. Air emissions from an anaerobic digestion system

Feedstock	Input, dry basis (lb × 10^6/d)	Sulfur emissions (lb SO_2/Btu × 10^6 output)	Nitrogen emissions (lb NO_2/Btu × 10^6 output)
Fresh cattle manure	2.20	0.108 (240)[a]	0.161 (218)[b]
Cattle manure, 5-month collection cycle	1.41	0.22 (152)[a]	0.187 (76)[b]
Fresh cattle manure and wheat residue	2.2 + 1-1	0.148 (334)[a]	0.171 (235)[b]

[a]Pounds of sulfur per day.

[b]Pounds of nitrogen per day.

Reprinted by permission of Kuo, T. M. C., and J. L. Jones, 1978. "Environmental and Energy Output Analysis for the Bioconversion of Agricultural Residues to Methane," pp. 143—76 in *Proc. of the IGT Conf. on Energy from Biomass and Wastes, Aug. 14—18, 1978, Washington, D.C.*, Institute of Gas Technology, Chicago, Ill.

4.4.5 Unspecified Emissions

Sulfur in fuel or fuel-producing feedstock is generally environmentally significant. Under oxidizing conditions sulfur is often converted to SO_2, and under reducing conditions it becomes H_2S, both of which present a possible air emission problem. When sulfur is scrubbed it turns into a sulfur-bearing sludge, which can become either a water-quality or solid-waste-disposal problem. Fortunately, except for manure and kelp, most biomass is relatively free of sulfur. The concentrations are about 0.3% in dry manure (Sect. 4.1.7, Table 4.11) and about 1% in dry kelp (Sect. 4.1.8, Table 4.13). Because the minimum-size mariculture farm will be 40,470 ha (100,000 acres) (Sect. 3.5.2) and will produce 90 dry metric tons of biomass per hectare per year (40 dry tons/acre/year), about 36,300 metric tons of sulfur will be released annually as a result of kelp conversion. Assuming that a 40,470-ha kelp system will yield 2.93×10^{16} J (2.78×10^{13} Btu) of methane annually (Wise et al. 1976), 1.25 kg of sulfur per 10^9 J (2.9 lb/10^6 Btu) will be released. Little is known about the form in which the sulfur will appear, although under anaerobic fermentation conditions, some of it may be converted to H_2S (Sect. 6.3.1).

4.5 Description of Full-Scale Conversion

A full-scale energy facility is readily perceived as a 1000-MW(e) power plant or its equivalent. However, a wood-burning 1000-MW biomass facility would require 14,500 metric tons/day of wood chips (3000 20-cm-diam trees per hour). Even the largest biomass processing facility (a paper mill) undertakes a formidable task in producing 2550 metric tons of paper per day (Beall et al. 1974). Thus the concept of *full scale* for biomass departs from the conventional technological approach in which full scale is almost always large scale.

Because the biomass resource is smaller than some of the nonrenewable resources and because it is regional, varied, and dispersed, the full-scale energy technologies that arise from its utilization are smaller and more regional, varied, and dispersed. Thus, to describe a full-scale facility is to describe a broad range of technologies covering many feedstock materials that are often applicable only to utilization at specific sites. This section presents this kind of description of full-scale biomass facilities, many of which are still in the conceptual design stage. The discussion is intended to accompany the analysis of potential environmental impacts of full-scale conversion facilities in Chap. 6.

4.5.1 Combustion

Table 4.27 is an overview of the effluent streams and land requirements for full-scale biomass combustion. Clearly, few emissions have been characterized with certainty, although some air emission studies have been conducted, primarily at power plants in which coal-fired boilers have been converted to a wood or wood-and-oil feedstock (Sect. 4.5.1.1).

4.5.1.1 Large-scale combustion. Except for minor differences, the technology for a large-scale wood-fired commercial facility such as a power plant is almost a duplicate of a coal-fired unit of the same capacity. In fact, it is reported that the substitution of wood firing in most existing coal-fired boilers is accomplished with only minor modifications (Hall et al. 1976). Additionally, air emissions for wood-fueled industrial systems are estimated as comparable with, or less than, those for systems using alternative fuels such as physically cleaned coal or low-sulfur western coals, the primary pollutant from wood combustion being particulate matter (Blake and Salo 1977).

In one study of a full-scale power station (Garabedian and Sanborn 1978), prior to conversion to burning wood, particulate emissions from co-firing 55% coal with 45% oil were 35.7 g/10^9 J input (0.083 lb/10^6 Btu). After conversion, particulate emissions increased more than threefold to 130 g/10^9 J input (0.302 lb/10^6 Btu), whereas the boiler was derated from 10 to 8.5 MW. The increased emissions were attributed to greater production of fine particles (less than 10 μm) with wood combustion and to the low efficiency of cyclone separators in removing fine particles. Gikis et al. (1978) believe that a large plant with emission-control equipment designed to capture particulates produced by wood firing could meet the Environmental Protection Agency's new source performance standards for particulates

Table 4.27. Overview of effluent streams and land requirements for large-scale wood combustion

	Biomass storage	Handling (pretreatment drying, etc.)	Combustion[a] (lb/Btu X 10^6 output)
Air emissions	Volatile HC gases (unidentified)	Volatile hydrocarbons released during drying (unidentified)	NO_x, 0.7; SO_2, 0; particulates, 0.1;[b] particulates, 8—10;[c] NC, 0.25; CO, 0.25
Aqueous emissions	Storage pile leachate (medium-high BOD not characterized)	Possible high BOD condensate from chip drying	None
Solid residuals	Materials aged beyond usability	None	Bottom ash about 1.39 lb/Btu X 10^6 input
Land requirements	Chip pile requires about three times as much space as equivalent coal pile		Less than 0.001 acre/Btu X 10^9/year output

[a]From Gikis, B. J., et al., 1978. *Preliminary Environmental Assessment of Energy Conversion Processes for Agricultural and Forest Product Residues,* EPA 600/7-78-047, NTIS PB-281 189, Municipal Environmental Research Laboratory, Cincinnati, Ohio.

[b]With controls.

[c]Without controls.

of 0.1 lb/10^6 Btu (~43 g/10^9 J) output. However, the importance of optimizing the combustion condition to control carbon monoxide emissions is often overlooked.

In 1978 a study was conducted of an operating power plant for which one coal-fired boiler, powering a 10-MW electric generator, had been modified to burn wood chips augmented with 20% No. 2 fuel oil (Golembiewski 1979). A layout of the plant and storage yard is given in Fig. 4.8. Wood chips are moved continuously from the outdoor storage area to a storage bin from which they are transferred to four gravity-fed chutes. The chips that are blown into the boiler land on a horizontal traveling grate which is supplied with underfire air. The No. 2 fuel oil, along with overfire air, is introduced above the grate bed from both sides of the firebox. As the grate and wood move along, the burning wood is converted to ash, which is discharged into a hopper at the end of the grate. The feed rate is 2.27 kg/s (9 tons/h) of 40% moisture wood and 0.66 m^3/s (175 gal/h) of oil (82% wood chips and 18% fuel oil on a Btu basis), producing 11.34 kg/s (90,000 lb/h) of steam. Firing oil above the glowing bed of embers would seem to be an effective way of igniting the distilled volatiles and the carbon monoxide above the bed.

The flue gases were ducted to an emission-control system composed of two high-efficiency mechanical collectors in series. Samples were taken of the flue gases, both at the inlet and the outlet of the mechanical collectors, to distinguish between the uncontrolled and controlled emission rates. On the basis of heat input, the uncontrolled particulate emissions averaged 1.47 kg/10^9 J (3.43 lb/10^6 Btu), whereas controlled emissions were 0.07 kg/10^9 J (0.17 lb/10^6 Btu). Particulate analyses indicated that the elements emitted at concentrations greater than 10 μg/dscm were Pb, Ba, Sr, As, Ga, Zn, Cu, Fe, Mn, Ti, and P. The number of small particles (<2.6 μm) appeared to increase in the controlled gas stream compared with those in the uncontrolled gas stream. Examination of both the collected fly ash and bottom ash for PAHs indicated the presence of phenanthrene at an average concentration of 0.89 μg/g in the bottom ash. A number of PAHs were found in the collector ash, including phenanthrene that was detected in one sample at 10 μg/g (Table 4.28). Examination of a comparable sample

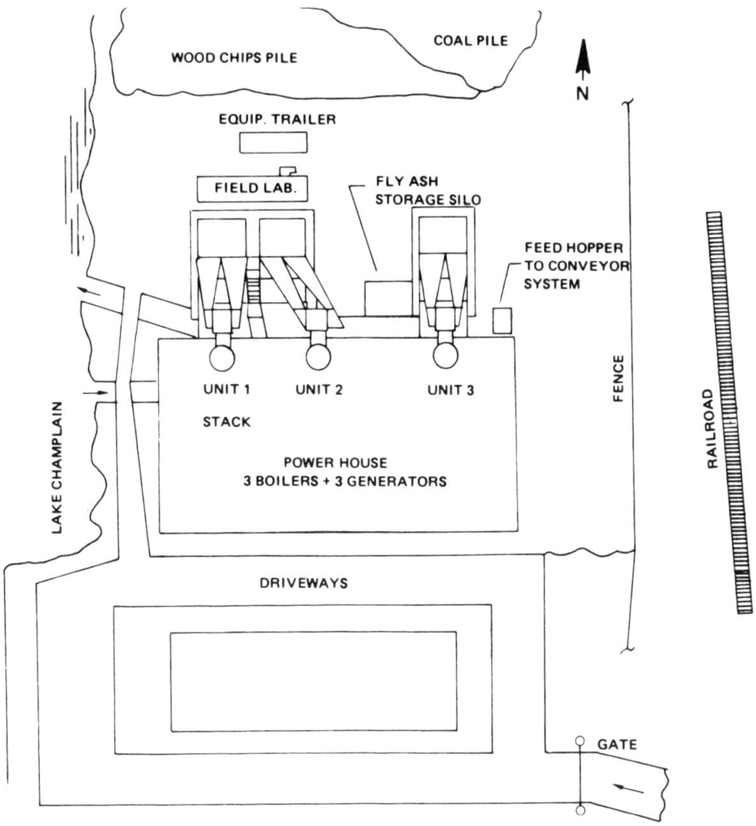

Fig. 4.8. Layout of the Burlington Electric Plant. Source: Golembiewski, M., 1979. *Environmental Assessment of Waste-to-Energy Process Wood and Oil-Fired Power Boiler*, Midwest Research Institute, Kansas City, Mo.

Table 4.28. Analysis of secondary collector ash for polycyclic aromatic hydrocarbons by gas chromatography—mass spectroscopy (μg/g)

Compound	Detection limit[a]	Run 1[b]	Run 1 duplicate	Run 2[b]
Acenaphthylene	0.56[c]	2.2	2.5	
Fluorene	0.54[c]			
Phenanthrene	0.24[c]	6.0	3.5	10
Anthracene	0.15[c]			
Fluoranthene	0.17[c]	1.6	0.62	3.5
Pyrene	0.16[c]	0.9	0.32	1.0
Chrysene	0.50[d]			
1,2-Benzanthracene	0.47[c]			<0.47[e]
Benzo[*a*]pyrene	6.7[c]			
Perylene	0.56[d]			
Indeno[1,2,3-*c*,*d*]pyrene	1.5[d]			
1,2,5,6-Dibenzanthracene	2.2[d]			
1,12-Benzoperylene	1.4[d]			

[a]Assuming a 20-g sample weight and a 1000-counter minimum signal.
[b]Values corrected to D-10 anthracene internal standard.
[c]Based on response factors determined by using 1 μl of a 1-ng/μl standard.
[d]Based on response factors determined by using 1 μl of a 5-ng/μl standard.
[e]Compound identity could not be verified by gas chromatography—mass spectrometry.

Source: Golembiewski, M., 1979. *Environmental Assessment of a Waste-to-Energy Process Wood and Oil Fired Power Boiler*, Midwest Research Institute, Kansas City, Mo

produced during firing of wood alone would have been instructive because little data exist to indicate whether the PAHs are derived from combustion of the wood, the oil, or a combination of the two. The 2.27-kg/s (9-ton/h) wood feed produced about 0.025 kg/s (200 lb/h) of bottom ash and an equal or somewhat greater quantity of fly ash. The residual ash was removed to an outdoor storage silo. If the wood ash is used as a soil supplement, as is common, it will be important to evaluate the disposition of the PAHs in the soil; if the ash is disposed of in a landfill, interaction of the PAHs with the subsoil environment and groundwater will be important.

The composition of the boiler exhaust is given in Table 4.29. The higher-than-expected SO_2 levels are considered to result from contamination of the wood feedstock by high-sulfur coal dust from the storage yard because the amount of sulfur in the wood (0.35%) was insufficient to produce the observed emissions. However, the high levels of carbon monoxide were difficult to explain in view of the overfire air introduced with the fuel oil. High carbon monoxide levels are symptomatic of either incomplete combustion or reduction of CO_2, a process that can occur whenever wood is fired on a fixed bed with an excess of underfire air (Sect. 4.2.1). The reported 245 ppm of carbon monoxide leads to an emission rate of 195,000 kg/year (215 tons/year) for the 10-MW unit—almost twice the reported emission rate of 490,000 kg/year (540 tons/year) to be expected

Table 4.29. Boiler exhaust gas composition[a]

	7/24 (11:45-17:00)		7/25 (09:00-16:00)	
	Average	Range	Average	Range
O_2, %	13.0	12.6-13.3	12.1	11.4-12.6
CO_2, %	7.9	7.9	7.3	6.8-7.9
SO_2,[b] ppm	155	110-230	120	90-160
CO, ppm	180	135-225	245	125-400
THC,[c] ppm	7	2-13	10	4-21
NO_x, ppm	68	60-77	64	57-71

[a] Gas samples were drawn from the duct downstream from the mechanical collectors and were analyzed by continuous monitoring equipment. Averages were computed from data points transcribed from the strip chart records at 15-min intervals.

[b] H_2O interference contributed about 20 ppm to all SO_2 readings. Values shown have been corrected for this discrepancy.

[c] Total hydrocarbon results were reported as propane.

Source: Golembiewski, M., 1979. *Environmental Assessment of a Waste-to-Energy Process Wood and Oil Fired Power Boiler*, Midwest Research Institute, Kansas City, Mo.

for a 50-MW wood-fired power plant (Hall et al. 1976). An analysis of the air quality implications of this emission rate is given in Sect. 6.1.

4.5.1.2 Medium-scale combustion. Highly efficient designs are now available for institutional wood-burning furnaces. One such unit heats the Agricultural Engineering Building at the University of Maine, Orono (Biomass Energy Institute 1978).

4.5.1.3 Small-scale combustion. Even though home wood-burning stoves are becoming more common, knowledge about their properties, construction, efficiency of operation, and pollution potential is still scarce. Wood stoves come in a wide variety of sizes, shapes, and designs, but most can be classified within the four types described below, depending on the amount of control applied to the combustion air (Wortman 1978).

Fireplaces. Fireplaces allow no control of combustion air (Table 4.30) and thus have a very low efficiency when operating (~13.6%). If the damper is left open when the fireplace is not in use, as it often is overnight, the fireplace will have a negative efficiency. One study showed that the net overall efficiency, as determined by the amount of furnace fuel burned during fireplace use, for a fireplace in a home with a thermostatically controlled furnace was zero, at best, when the fireplace was covered at night and negative otherwise (Shelton 1978). The increased ventilation of cold air provoked by use of a fireplace is a major contributor to nullifying its heating effect.

In addition to low efficiency, a fireplace has a high emission rate for particulates, carbon monoxide, and hydrocarbons (Sect. 4.4.1). Despite the zero or negative efficiency and high pollution potential, fireplaces undoubtedly have a high, overriding aesthetic appeal.

Nonairtight stoves. These are freestanding stoves with loose-fitting doors through which air can leak readily into the firebox. They include the well-known "pot belly" stove as well as the ordinary Franklin stove. Unlike the fireplace, which heats primarily by direct radiant energy from the flames and coals, these stoves heat by radiant energy emitted from the stove walls, chimney, and firebox, as well as by convection of heated room air in contact with the hot stove walls. The relative amounts of energy transferred by the two

Table 4.30. Rough estimate of fireplace energy flows

	Wood energy consumption (%)	Power (Btu/h)
Wood consumption, 20 lb/h (8600 Btu/lb)	100	172,000
Direct or primary radiation	10	17,200
Indirect or secondary radiation	5	8,600
Chimney conduction contribution— 10% of sensible heat entering chimney for a two-story house and an interior chimney with exposed sides	7	12,000
Air heat loss 1500 lb/h, 40°F colder outdoors than indoors	−8.4	−14,500
Net efficiency 10 + 5 + 7 − 8.4 = 13.6% while operating (not including overnight losses):		
Net heat gain 17,200 +8,600 +12,000 −14,500 = 23,300 Btu/h		

Reprinted by permission of Shelton, J., 1978. *The Woodburner's Encyclopedia,* Vermond Crossroads Press, Waitsfield, Vt.

mechanisms are given in Table 4.31. For most stoves, more energy is transferred by radiation than by convection (as in a fireplace), but the aesthetic pleasure of seeing the dancing flames and glowing coals is absent.

Like a fireplace, a stove that draws air from the house to supply the necessary oxygen for combustion pulls outside air into the house and discharges heated house air into the chimney. In a nonairtight stove, the air flow is not controllable, and thus the rate of combustion will not be controllable. The fuel will burn rapidly, inefficiently, and with a hot flame; the fast flowing, very hot combustion gases will be vented to the outside. The rapid inefficient combustion causes particulates, carbon monoxide, and hydrocarbons to be produced and vented (Sect. 4.4.1).

Nearly airtight stoves. An example of a stove that is nearly airtight is the Franklin-

Table 4.31. Energy transferred by radiation and by convection from an exposed hot surface, such as a radiant stove, to its surroundings

Temperature of surface (°F)	Total energy transferred ($Btu/h/ft^2$)	Natural convection (% of total)	Radiation (% of total)
80	15	28	72
100	51	35	65
150	168	39	61
200	315	40	60
400	1230	35	65
600	2850	29	71
800	5430	23	72
1000	9370	18	82
1200	15100	15	85
1400	23100	12	88

Note: The total heating rate per square foot of stove surface area increases dramatically as the surface becomes hotter. At all temperatures, more than one-half of the energy is in the form of infrared radiation. (In circulating stoves, air is heated as it circulates through interior passages; in this case, the convected hot air carries more heat into the room than does radiation.)

Reprinted by permission of Shelton, J., 1978. *The Woodburner's Encyclopedia,* Vermont Crossroads Press, Waitsfield, Vt.

type stove with welded seams. This type of stove has an essentially airtight body with tightly fitted, but nongasketed, doors. Nearly airtight stoves can be damped to kill the flames but not the burn. Thus, the fuel will burn for a longer time at a lower rate. Additionally, the stoves draw so little air, compared with the normal air exchange rate for a house, that they do not appreciably add to the heating needs of the house as can a fireplace or nonairtight stove (Table 4.32). The emission rates for nearly airtight stoves have not been reported, but they are expected to be less than those from fireplaces and nonairtight stoves.

Airtight stoves. This type of stove has an airtight body and close-fitting gasketed doors. Known as circulating stoves and space heaters, airtight stoves use convection rather than radiation as the dominant mechanism for transferring heat to the room. Either natural convection (Fig. 4.9) or small blowers are employed to move warm air into the room. There is no fundamental reason for a circulating airtight stove to have a higher efficiency than a radiant airtight stove; the choice depends on the sensation of thermal comfort (Shelton 1978). The range of efficiencies covered by all kinds of airtight stoves is reported as probably between 40 and 65%. This efficiency is compared with advertised new-furnace efficiencies of 80 to 85% (which are thought to be probably closer to 75% in use) (Shelton 1978). If wood burning is to become an effective home heating method, studies will be needed to determine designs that combine combustion efficiency with optimal heat transfer and emissions control.

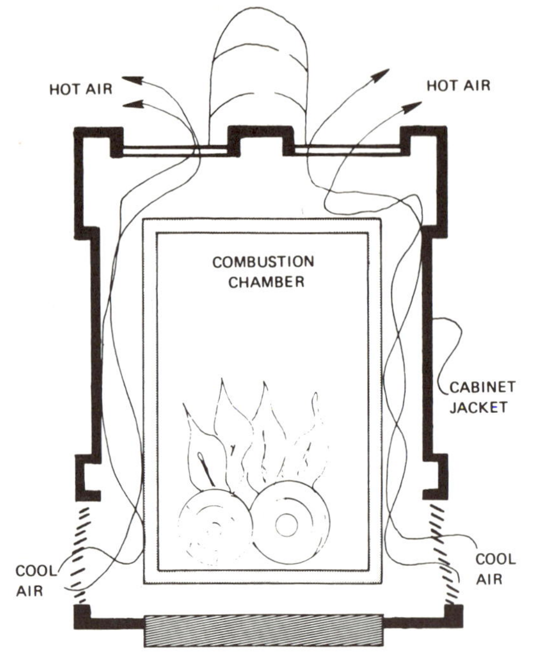

Fig. 4.9. The Ashley Imperial (model C-60), a circulating stove. Source: Shelton, J., 1978. *The Woodburner's Encyclopedia*, Vermont Crossroads Press, Waitsfield, Vt.

4.5.2 Thermochemical Conversion

Table 4.33 is a summary of the waste streams and land requirements to be expected from three sizes of thermochemical biomass processing facilities. Details of the processes are given in Sects. 4.5.2.1 to 4.5.2.3.

4.5.2.1 Large-scale thermochemical conversion. In 1978 Bechtel National Corpor-

Table 4.32. Typical air consumption of wood heaters and air exchange rate of houses

	Air consumption (lb/h)	Air exchange rate (ft^3/min at 70°F)
Small closed stove	100–200	22–44
Large closed stove	150–300	33–67
Open stove (e.g., Franklin)	250–500	56–110
Fireplace (with windows and doors closed)	400–1500	90–330
Typical house ventilation rate	500–2000	125–500

Note: Because most of the mass flow up chimneys is air, these numbers also represent typical flue-gas flows. (Data from various sources, including the author's measurements.)

Reprinted by permission of Shelton, J., 1978. *The Woodburner's Encyclopedia*, Vermont Crossroads Press, Waitsfield, Vt.

Table 4.33. Details of waste streams and land requirements — thermochemical conversion

	Large-scale (Bechtel conceptual design)	Medium-scale (Garrett Engineering conceptual design)	Small-scale (TVA—Maryville mobile pyrolysis unit)
Input	1000 wet tons/d manure	100 wet tons/d manure	3 tons/h wood waste
Product output	8986 × 10^6 kJ/d (8500 × 10^6 Btu/d); gas: 25 × 10^6 scfd at 340 Btu/scf	897 × 10^6 kJ/d (850 × 10^6 Btu/d); gas: 1.7 × 10^6 scfd at 500 Btu/scf	940 × 10^6 kJ/d; total: 891 × 10^6 Btu/d; gas: 8 × 10^6 Btu/d charcoal: 24 tons at Btu/lb = 528 × 10^6 Btu/d; 11,000 pyrolysis oil: 2400 gal at 148,000 Btu/gal = 355 × 10^6 Btu/d
Air emissions	4250 scfm, mostly CO_2, trace hydrocarbons, 10 ppm H_2S (6 lb/d H_2S	?	?
Wastewater	12,360 liters/d	?	None; vapor goes with produced gas
Solids	403 metric tons/d char and ash (443 tons/d) 9 metric tons/d sulfur (10 tons/d)	24 metric tons/d ash (26 tons/d)	Ash generated offsite when charcoal is used
Land	6 ha	<1 ha	<1 ha

ation completed a conceptual design for a commercial-sized gasification plant based on Texas Technological University's experience with using feedlot manure to produce synthetic gas on a pilot-plant scale (Hipkin and Basuino 1978). The Texas Technological University process is well suited for insight into commercial biomass gasification. The process has operated on a pilot scale with at least two feedstocks, manure and wood chips; the product gas is suitable for direct combustion, for use in ammonia manufacturing, or for conversion to methanol. The pilot process incorporates both a gas cleanup system and a tar collection train. Additionally, under some operating conditions with manure, the gas contains up to 5 mole % ethylene, a chemical that can be recovered as a valuable by-product.

Manure feed and product descriptions for the commercial design are given in Table 4.34. A

Table 4.34. Manure feed and product description, syngas option

Component	Manure feed[a]	Syngas product	Sulfur by-product
H_2, moles/h		1,438.5	
N_2		905.4	
CO		502.3	
CH_4		307.4	
C_2H_6		36.7	
C_2H_4		148.7	
Acetylenes and diolefins		24.9	
H_2O		8.1	
S			26.6
Total moles/h		3,372.0	26.6
Total lb/h	130,750	53,619	850
Scfm[b]		20,180	
LHV,[c] Btu/scf		340	

[a]Manure feed: 65,370 lb/h miscellaneous organics, 32,690 lb/h ash, and 32,690 lb/h water, as received (Texas Tech basis).
[b]Standard conditions at 32°F, 14.7 psia.
[c]LHV = lower heating value.
Source: Hipkin, H. G., and D. J. Basuino, 1978. *Syngas from Manure — A Conceptual Plant Design*, Bechtel National, Inc., Research and Engineering, San Francisco, Calif.

nominal feed of 907,000 dry kg/d (1000 tons/d) produces about 708,000 m^3/d (25 million scf/d) of gas with a low heating value of 12,670 kJ/scm (340 Btu/scf). The design calls for a plant site of approximately 6 ha (15 acres). As with all commercial biomass conversion facilities, a large feedstock storage area will be required (Fig. 4.10). In this case it amounts to 0.6 ha (1.5 acres), or 10% of the plant site, and will store 1300 metric tons of manure. A site-specific environmental analysis of the facility would include consideration of the long-term effect of runoff from the manure storage area (see Sects. 2.3.5 and 4.1.7). Rain falling at the rate of 1 in./h on the manure storage pad will necessitate a drainage system designed to handle a water flow of 160 m^3/h (42,000 gal/h to an American Petroleum Institute separator, in which insoluble oils are separated and the remaining stream is discharged to a municipal sewage system. The commercial design has no provision for controlling odors that may emanate not only from the raw manure pile but also from gaseous discharge of organics and possibly from the ash and char (Sect. 6.3).

Biomass feedstocks other than manure will also require onsite storage. With such storage the question of feedstock runoff, which may at first appear to be less pressing for feedstocks other than manure because of the absence of odor, will need to be addressed. For example, little is known about the runoff characteristics of wood chips during open storage except that the runoff may be slightly acidic (Sect. 4.1.2.1). Prevention of accidental ignition of wood storage piles will be important; if prevention is accomplished by periodic wetting of the pile, runoff may be a problem. On the other hand, fire not only destroys the feedstock, it may be a source of unplanned environmental emissions (Sects. 4.4.1 and 4.4.2.1).

The heart of the syngas process is the reactor section, in which a crude synthesis gas is produced (Hipkin and Basuino 1978). Manure

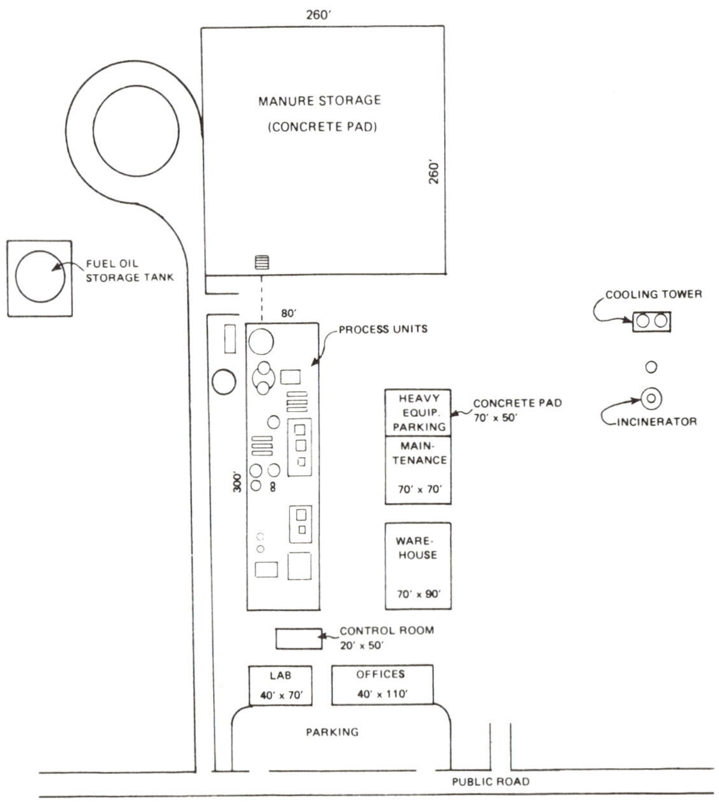

Fig. 4.10. Plant site layout. Source: Hipkin, H. G. and D. J. Basuino, 1978. *Syngas from Manure—A Conceptual Plant Design*, Bechtel National, Inc., Research and Engineering, San Francisco, Calif.

is added to the top of the hot reactor, where it falls countercurrent to a stream of steam and air introduced through a distribution plate at the bottom. The crude syngas leaves the reactor laden with tar, particulates, and water vapor, and a char residue is discharged from the bottom of the reactor. The gases then pass through cyclones in which a major portion of the particulates are removed and returned to the bed. The remaining particulates in the gas (the "fines") are removed by electrostatic precipitation; the tar is removed by quenching the gas in a condenser and then scrubbing it with quench water. The tar and water are separated; the tar is recycled to the reactor and the water is pumped to aqueous waste. The wastewater will undoubtedly contain a high concentration of soluble organic compounds extracted from the tars. Such aqueous waste streams are likely to have a high BOD, perhaps ranging from 500 to 3000 mg/liter (DiNovo et al. 1978). For a manure feedstock the effluent is likely to contain primarily long-chain fatty acids (Beck 1979), whereas waste-water from scrubbing wood-derived tars is more likely to contain aldehydes, ketones, alcohols, phenols, PAHs, and a small amount of free fatty acids (DiNovo et al. 1978).

After particulate removal the gas enters a gas-purification unit in which it is scrubbed to remove CO_2 and H_2S. A Stretford unit processes the CO_2 and H_2S, producing 9000 kg/d (10 tons/d) of solid sulfur and 120 scmm (4250 scfm) of a waste gas containing mostly CO_2, with a small amount of hydrocarbons, and 10 ppm H_2S. This gas is discharged to the atmosphere. Additionally, 402,000 kg/d (443 tons/d) of ash and char are discharged to a landfill. According to the design bases, the feedstock manure will be trucked to the site at the rate of one truckload every 10 to 12 min; solid waste will be removed at a frequency of one truckload every half hour.

According to the design specifications, aqueous waste, produced at the rate of 0.5 m^3/min (136 gal/min), will be treated by incineration because of the difficulty of treating the highly organic condensate biologically. Incineration is energy-intensive but is apparently justified by the difficulty of managing the aqueous waste. More detailed knowledge about the precise composition of the waste streams may make other options feasible.

4.5.2.2 Medium-scale thermochemical conversion. Because manure for conversion to gas will be produced primarily in beef cattle feedlots and because transportation of the manure is costly, it is likely that a gasification facility will be located near the feedlot. Garrett (1977*b*) evaluated both the number of feedlots of various sizes in the United States and the economics of gasification plants of various sizes. He showed that a small (1.8 metric ton/d) plant would be prohibitively expensive and that a very large (635 metric ton/d) plant would require feedstock from several feedlots. A 90 metric ton/d plant supplied by a feedlot containing 36,000 head of cattle was considered a reasonable size for process analysis.

Manure is delivered to the plant fairly regularly and is stored on a 557-m^2 (6000-ft^2) concrete pad with walls 1.8 m (6 ft) high on three sides. The manure is fed into a screw feeder that is heated with low-temperature flue gas produced by char combustion. In heating the feedstock the flue gas is cooled, producing a flue gas condensate and a vented flue gas. The conveyor is partially evacuated, causing the moisture in the biomass to vaporize and to be drawn off and condensed, thereby producing another condensate. Knowledge of the composition of both the condensates and the vented flue gas is important for evaluating the pollution potential of this pretreatment stage.

In the reactor the biomass passes through a dryer into the pyrolysis unit at 649 to 760°C (1200 to 1400°F), where it is mixed with hot char and water-gas reaction products. Pyrolysis gas leaving the pyrolyzer enters the gas purification system, where it is scrubbed to remove CO_2 and H_2S. The gas is subsequently cooled and compressed, producing an aqueous condensate for disposal.

The 90 metric ton/d plant will produce 0.05 million scm (1.7 million scf) of 527 kJ (500 Btu) gas per day for sale and 23.6 metric tons of ash per day for disposal. All electricity was assumed to be generated onsite from the high-pressure steam produced by combustion of pyrolysis char.

4.5.2.3 Small-scale thermochemical conversion. A prototype small-scale biomass converter has been tested by the TVA since

January 1979. The transportable pyrolysis unit is mounted on a 10-m flatbed trailer (Fig. 4.11) and consists of three majors parts: (1) a pyrolysis chamber in which wood-waste chips are charred by hot recirculating pyrolysis gases, driving off hydrocarbon volatiles and gas; (2) a condenser unit in which hydrocarbons are condensed from the gas stream, producing pyrolysis oils; and (3) a combustion chamber in which a portion of the pyrolysis gas is burned to provide heat for the pyrolysis reaction. No air is admitted to the pyrolysis chamber, so little nitrogen should be present in the fuel gas (a small amount of air is introduced along with the feedstock). The reactor is part of an internal loop in which a portion of the pyrolysis gas is fed through a heat exchanger. In the heat exchanger the gas picks up combustion heat and carries it back to the reactor. The nonrecirculated gas stream is sent through the condenser and is then divided; part of it goes to an external heating plant and part of it goes to the combustion chamber.

The unit is rated at 2720 kg/h (3 tons/h) of dry wood waste and is expected to produce 970 kg (1 ton) of charcoal, 0.38 m^3 (100 gal) of pyrolysis oil, and 8.4 million kJ (8 million Btu) of fuel gas. Although the charcoal is considered an environmentally clean product, the charcoal quench water should be investigated. If the oil is condensed at a temperature well above the boiling point of water, there should be no aqueous component in the oil and thus no aqueous effluent. The gas is dirty, containing fines, charcoal, and hydrocarbon particulate. However, it will be immediately and completely incinerated in a boiler. No emissions of environmental concern are expected from the unit, although the content of the flue gases will be monitored.

Comparably sized crop-residue gasifiers have been reported (Hodam and Williams 1978). Measured stack emissions from the gasification of wood waste were 129.7 ppm for NO_x and 0.319 kg/h (0.703 lb/h) for particulates. In addition to the stack emissions, smoke leakage was reported from the gasification vessel, and airborne dust was observed during char ash collection. Condensate tar, which collected on cooler surfaces, was a black unidentified liquid produced at the rate of 19 to 30 liters (5 to 8 gal) during startup and 0.95 liter/h (1 qt/h) during operation. The tar consisted of light and heavy fractions containing the metals listed in Table 4.35.

Table 4.35. Metals contained in gasifier tar (ppm)

	Run 64	Run 66
Manganese	34	11
Iron	8700	927
Zinc	94	9
Lead	2	1.5
Cadmium	30	45

Reprinted by permission of Hodam, R. H., and R. O. Williams, 1978. "Small Scale Gasification of Biomass to Produce a Low Btu Gas," in *Proc. of the IGT Conference, Energy from Biomass and Wastes, Aug. 14—18, 1978, Washington, D.C.*, Institute of Gas Technology, Chicago, Ill.

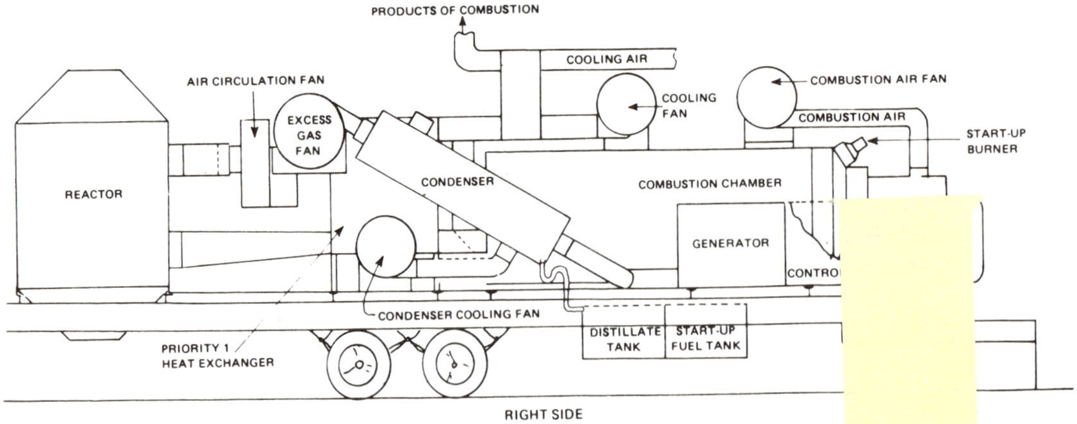

Fig. 4.11. Mobile wood-waste pyrolysis unit.

4.5.3 Anaerobic Digestion

During anaerobic digestion, also known as methane fermentation, wet biomass materials are converted microbially in the absence of oxygen to methane gas, carbon dioxide, and other impurity gases. Anaerobic digestion employs known technology, which is practiced commercially on a large scale in the conversion of municipal sewage wastes. Although the process chemistry is not completely understood, operating manure conversion facilities currently vary from the home-size biogas generator (common in India and China) to near-commercial-size facilities such as that in Bartow, Florida (Wise et al. 1978). Additionally, bioengineering is currently an area of active research in which new and innovative approaches applicable to feedstocks other than manure or municipal sewage are being proposed, tested, and demonstrated (Pitt 1979 and Genung et al. 1979).

4.5.3.1 Large-scale anaerobic digestion. The design criteria and flow diagram for a large, full-scale manure-to-methane conversion facility are shown in Table 4.36 and Fig. 4.12 (Varani et al. 1977). The facility comprises four major components: (1) a mixing unit containing a 15-kW (20-hp) mixer; (2) a digestion complex of four earthen basins 41 × 41 × 9 m (135 × 135 × 30 ft high); (3) a sludge-processing facility containing two solid bowl centrifuges, each with a capacity of 0.66 m^3/min (175 gal/min); and (4) algae ponds consisting of 2.4 ha (6 acres) of concrete basins 0.46 m (18 in.) deep. The facility will process about 31,750 kg (350 tons) of manure daily, will store a ten-day feedstock supply onsite, and will consume from 123,300 to 246,600 m^3 (100 to 200 acre-ft) of water per year. Water will be

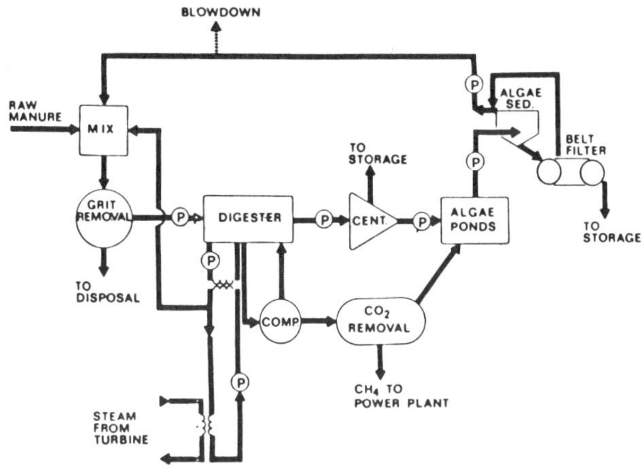

Fig. 4.12. Selected flow schematic for manure-methane facility. Source: Varani, F. T., et al., 1977. "The Design of a Large-Scale Manure/Methane Facility," in *Fuels and Energy from Renewable Resources*, D. A. Tillman et al., eds., Academic Press, New York.

recycled to the maximum extent, with blowdown only to prevent buildup of substances such as ammonia, sodium, calcium, and magnesium, which are toxic to the anaerobic organisms. The digester will operate at 35°C (95°F) and will be heated with waste heat from a nearby power plant. The digesters will be monitored from a control building for temperature, pH, alkalinity, volatile acids, gas production, and percentage of CO_2. Process sludge will be degassed by use of vacuum pumps and will be dewatered by use of centrifuges. The released gases will be discharged to the atmosphere, whereas the released water will be sent to algae ponds that supply recycle water. The dewatered sludge (cake), which will contain about 25% solids, will be stored onsite, where it is expected to dry to 50% solids. Six months of sludge storage will be provided

Table 4.36. Design criteria for full-scale manure-to-methane conversion facility

Number of cattle to supply manure (design point)	50,000
Annual manure load, tons/year	125,000
Total gas produced, ft^3 CH_4 at STP conditions, 24 h	1,226,450
Design plant loading rate, lb volatile solids/ft^3 digester	0.20
Volatile solids destruction, %	55
Digester temperature, °F	95

Reprinted by permission of Varani, F. T., et al., 1977. "The Design of a Large-Scale Manure/Methane Facility," in *Fuels and Energy from Renewable Resources*. D. A. Tillman et al., eds., Academic Press, New York.

onsite. Because 40 to 50% of the weight of the original feedstock is left as residuals (Miller 1976), storage space for 54 million kg (60,000 tons) of sludge will be required. The sludge will eventually be disposed of as fertilizer or as cattle feed supplement (see Sects. 4.1.7 and 6.3). The algae in the ponds will be harvested after a five-day detention, settled, and then concentrated to 5 to 6% solids. The algae are expected to remove toxic elements from the effluent.

Emissions from the plant to the environment could occur from disposal of the blowdown effluent; from transfer of the centrifuge centrate to the algae ponds, with possible subsequent drainage to groundwater; from outdoor storage of the sludge solids; and from disposal of the gas cleanup effluents. The final disposition of the algae is unclear. Capital costs of the facility have been estimated at between \$7.5 and \$10 million, with an annual operating budget between \$400,000 and \$700,000 (Varani et al. 1977).

4.5.3.2 Medium-scale anaerobic digestion. A full-scale, on-farm manure-to-methane facility has been operating since 1975 at a 400-head, 100-ha dairy farm in Monroe, Washington (Shrinde 1978). The anaerobic digester, which uses known municipal sewage treatment (MST) technology, consists of two 190-m^3 (50,000-gal) glass-enameled steel tank reactors that operate at 35°C (9521°F). The facility uses one tank and 200 cattle units of manure daily [one cattle unit = one 454-kg (1000-lb) animal], or 14.4 m^3 (3800 gal) of a slurry containing 9% solids, the highest viscosity that could be pumped with the sewage treatment equipment. For these specifications and with a 12-d hydraulic retention time, daily fuel gas production has been 200 m^3 (7000 ft^3), 30 to 35% of which was carbon dioxide. Manure is collected by flushing the loafing shed with water and forcing the slurry into sumps connected by pumps to an earthen storage lagoon (Fig. 4.13). Manure is applied to the fields from the lagoon by an irrigation system that uses a manure sprinkler gun.

The methane facility is composed primarily of digestion tanks, a mixing system, a heat exchanger, and a boiler (Skrinde 1978). The tanks are off-the-shelf manure storage tanks. The mixing system is a draft tube that has concentric walls and that also acts as a heat exchanger. Gas is pumped through pipes that extend to the bottom of a draft tube. The rising bubbles displace manure in the tube in a pumping-mixing action. A counterflow heat exchanger, designed to transfer 58% of the effluent heat to the influent fresh manure, has not functioned. Fuel gas to fire the boiler for heating the digester and the influent used about 35% of the total energy produced. The fuel gas was also burned in a gas-fired generator, which had been fitted with a special carburetor, giving an efficiency of conversion to electricity of only 11%. Little attention was given to the digester waste (pollution and odor control) or to cleanup of the H_2S in the fuel gas, although these were recognized as potential problems. It was acknowledged that there was little urgency among farmers to change existing practices to mitigate pollution. The actual fuel gas produced was flared or stored. Space heating, cooking, and hot water for two houses on the farm would have required gas from only 30 cows, leaving excess gas to compress, to bottle and sell, to convert to electricity at an extremely low efficiency, or to use in farm equipment converted to burn digester gas (see Chap. 7 for socioeconomic implications of on-farm energy production). The Monroe facility was constructed at a cost of \$119,000, but an effective system of the same size was judged to be possible at one-half that cost. This means that for a \$60,000 to \$120,000 capital investment, a farm can produce \$21.00 worth of gas daily, assuming \$3.00 per 1000 scf as a purchase price for pipeline gas.

4.5.3.3 Small-scale anaerobic digestion. The home-size, nonstirred biogas unit is common in both India and China, but the likelihood of its immediate acceptance in this country is questionable. A typical simple batch digester is shown in Fig. 4.14 (Chiranjivi 1978). The unit is circular with two concentric walls, one with a 1-m (3.5-ft) radius and the other with a 1.6-m (5.25-ft) radius. The space between the walls is filled with water and serves as a seal for the gasometer, which is a circular galvanized-iron tank open at the lower end. The digester is filled with about 3600 kg (4 tons) of dung and vegetable waste to 0.3 m (1 ft) below the top, and water is added to cover the mass. Vegetable matter prevents the formation of a tight mass of fibrous material over the top of the slurry, which would eventually inhibit gasification. The unit operates at ambient

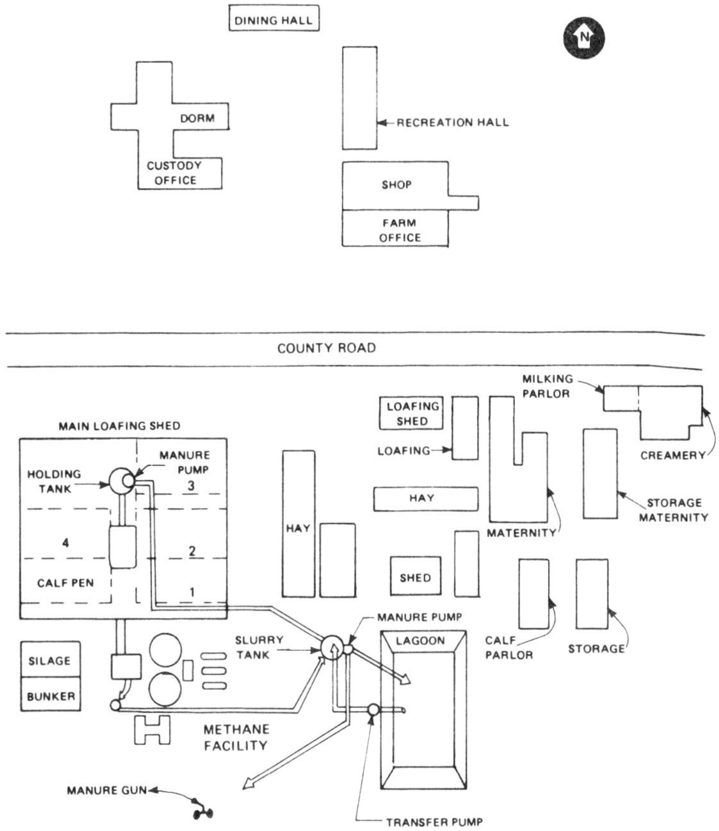

Fig. 4.13. Monroe State Dairy Farm plot plan. Source: Skrinde, R. T., 1978. *Operation of an Anaerobic Digester at the Washington State Dairy Farm,* Tracy & Brunstiom Co., Seattle, Wash.

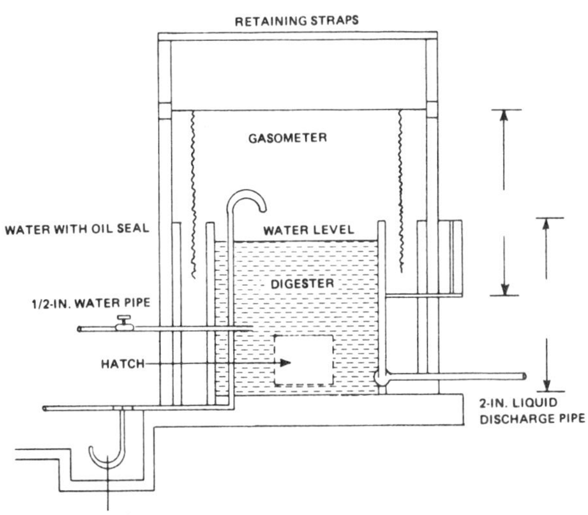

Fig. 4.14. Batch digester. Source: Chiranjivi, C., 1978. "Design Analysis of Small-Scale Anaerobic Digesters in India," in *Symposium on Energy from Biomass and Wastes, Washington, D.C., Aug. 14-18, 1978,* Institute of Gas Technology, Chicago, Ill.

temperature, producing 3.4 m^3 (120 ft^3) of gas per day for 30 to 40 consecutive days and dropping to 2.5 m^3 (90 ft^3) after 40 d. In freezing or cold weather, when the gas is needed most, activity is severely curtailed or stops entirely. After two months, the water is completely removed, the tank is cleaned, and the unit is recharged. At least two units are required for an uninterrupted gas supply.

4.5.3.4 Proposed systems. Systems for the conversion of aquatic biomass are still in the very early stages of development. Researchers have reported that the anaerobic digestion of each pound of algae in a batch reactor with a retention time of 30 d can produce about 0.23 m^3 (8 ft^3) of gas containing 70% methane (Golueke et al. 1957). However, processing algae-rich waters to produce methane in a continuous operating mode is yet to be demonstrated.

The ANFLOW fixed-film anaerobic bioreactor (Genung et al. 1979), which has been technically demonstrated in a 20-m^3/d municipal wastewater treatment pilot plant, has been proposed as a means of continuous methanation of microalgae, macroalgae, and water hyacinth (Pitt 1979). The ANFLOW system employs microbial films, immobilized on stationary packing material, through which dilute liquid wastes are passed in an upward flow. The continuous-flow process combines biophysical filtration with anaerobic fermentation and can accommodate feeds containing both suspended and soluble materials (Genung et al. 1979).

The process proposed by Pitt (1979) involves three bioreactors in series. The first reactor concentrates algae and produces in-situ solubilization. The upflow mode of operation concentrates algae and minimizes the customary algae dewatering costs. Solubilization is achieved by enzyme-producing bacteria maintained as fixed films on the solid support material. The effluent from the first column is forwarded to a second bioreactor that also employs fixed microbial films of acetogenic bacteria. The organic-acid-containing effluent from the second stage enters a third bioreactor for methanogenesis. Schemes for using algae ponds for the final disposition of biomass conversion wastewater (Sect. 4.5.3.1) are sufficiently prevalent to warrant further study of the algae conversion process as the terminal step of complete biomass-to-energy systems.

Laboratory studies of the anaerobic digestion of giant brown kelp, *Macrocystis pyrifera,* indicate that batch or semicontinuous processing of kelp is possible over extended periods without adding external nutrients (Klass 1977). In this work, many of the important parameters in kelp digestion were delineated, such as characteristics of the feed slurry as compared with those of the digester effluent (Table 4.37), but additional study is clearly needed if marine biomass is to be seriously considered as a fuel feedstock.

4.5.4 Fermentation

Fermentation is an ancient art that has become current technology and that is now being reexamined to establish its energy efficiency and environmental compatibility. Figures 4.15 and 4.16, flow diagrams of a conventional and a modified alcohol production scheme, respectively, exemplify this trend. Whereas the old production system produced copious solid wastes (such as bagasse and slops), an acid aqueous effluent, and uncontrolled air emissions, the new plant will be expected to recycle to the maximum extent, apparently with no waste stream and requiring no external source of energy.

4.5.4.1 Industrial-scale fermentation. In 1949, long before the present energy crisis, Prescott and Dunn (1949) predicted the current situation: "The subject of ethyl alcohol production by fermentation has assumed new interest on account of attempts to find substi-

Table 4.37. Comparison of feed kelp and digester effluent slurries for semicontinuous run

	Feed slurry	Digester effluent[a]
Volatile solids, wt %	2.88	1.66
pH	6.28	7.12
Alkalinity, mg/liter as $CaCO_3$	1,448	5,831
Ammonia nitrogen, mg/liter	16	547
Organic nitrogen, mg/liter	1,342	707
Chemical oxygen demand, mg/liter	33,012	18,368
Volatile acids (filtrate), mg/liter as HOAc	44	311
Specific conductivity, μmho/cm	22,200	32,133

[a]Mean values.

Reprinted by permission of Klass, D. L., 1977. "The Anaerobic Digestion of *Macrocystis pyrifera* under Mesophilic Conditions," Table 10, p. 345, in *Clean Fuels from Biomass Symposium*, Orlando, Fla., Jan. 25—28, 1977, Institute of Gas Technology, Chicago, Ill.

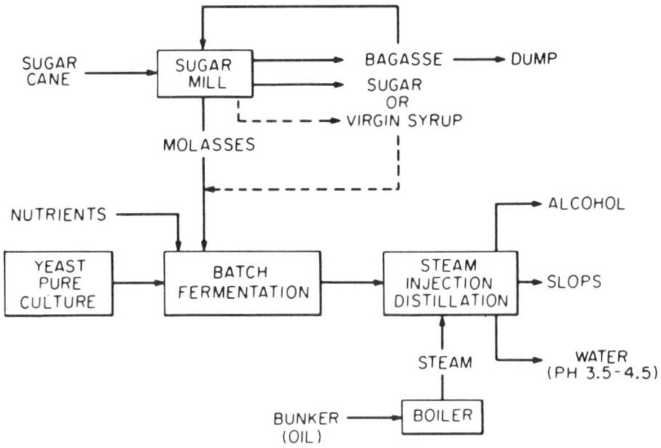

Fig. 4.15. Current alcohol production process. Source: Espinosa, R., et al., 1979. "Alternatives for Energy Savings at Plant Level for the Production of Alcohol for Use as Automotive Fuel," in *Biotechnology in Energy Production and Conservation,* C. D. Scott, ed., Biotechnology and Bioengineering Symposium 8, Interscience, New York. Copyright 1979, reprinted by permission of John Wiley & Sons, Inc.

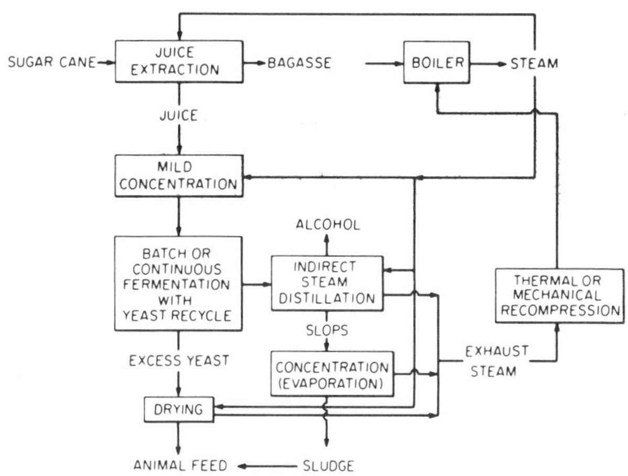

Fig. 4.16. Modified alcohol production process. Source: Espinosa, R., et al., 1979. "Alternatives for Energy Savings at Plant Level for the Production of Alcohol for Use as Automotive Fuel," in *Biotechnology in Energy Production and Conservation,* C. D. Scott, ed., Biotechnology and Bioengineering Symposium 8, Interscience, New York. Copyright 1979, reprinted by permission of John Wiley & Sons, Inc.

tutes for gasoline. Blends of alcohol with gasoline, especially a blend containing 10% ethyl alcohol, may be used satisfactorily in the present internal-combustion type of motor. Present-day demands for motor fuels are great, but the sources of petroleum are limited."

Ethyl alcohol may be produced from any fermentable sugar or from any other carbohydrate source that can be hydrolyzed into a fermentable sugar. Hydrolysis may be carried out chemically (acid hydrolysis) or enzymatically (enzyme hydrolysis). Raw materials accessible to industrial-scale alcohol fermentation in the United States fall into three classes: (1) the saccharine substances, such as sugarcane, sweet sorghum, sugar beets, and molasses; (2) the starchy substances, such as corn, wheat, and other grains; and (3) the cellulosic materials, such as wood, straw, or agricultural residues.

Sugarcane. Production of alcohol from sugarcane is by far the simplest process. Mechanically harvested cane is cleaned to remove leaves, debris, and dirt before it enters the milling phase. Some cane is cleaned dry, but when wet cleaning is practiced, large quantities of water are used (J. E. Atchison Consultants, Inc. 1977). The clean cane is then cut into small pieces, shredded by a device such as a hammer mill, and passed through roll presses to further break up the cane and press the juice out. From 92 to 95% of the sugar is extracted from the cane as a strong juice, leaving a residual material called bagasse (Sect. 4.1.4), which contains about 50% moisture and some residual sugar. From 97 to 98% of the sugar can be removed by passing the rolled cane particles through a continuous diffuser, in which the sugar juice is pumped countercurrent to the flow of cane (J. E. Atchison Consulants, Inc. 1977). After leaving the diffuser, the cane particles again pass through rollers, where the bagasse is squeezed to 50% dry. Twenty-five weight percent of the sugarcane ends up as bagasse; if discarded, it becomes a solid waste disposal problem, and if burned in the mill boilers without particulate control systems, it becomes an air-quality problem (Sect. 4.4.1).

Prior to fermentation the raw mash is diluted to an optimum sugar concentration of about 10% and an optimum pH of 4.0 to 4.5 and then heated to a temperature of 20 to 26°C. A yeast starter is added and thoroughly mixed with the mash in the fermenter, where the mixture "works" for about 40 h. The fermented mash, referred to as "beer," is distilled in a continuous still to separate the ethanol and fusel oil from the "slops," or stillage. The high-alcohol fractions (60 to 90% ethanol) then are concentrated by additional distillation to 95% ethanol (190 proof), whereas the low-alcohol fractions are redistilled with the next batch of beer. The stillage, which amounts to almost as much mass as the original input of sugarcane, is the largest-volume by-product of alcohol fermentation (Yang and Trinidade 1978). In addition, discarded, untreated stillage has a high pollution potential. Table 4.38 gives the material balance for a typical Brazilian sugarcane distillery, using 1000 liters (264 gal) of anhydrous alcohol as a basis. A typical distillery produces 150,000 liters/d (about 40,000 gal/d) of alcohol and 12 to 13 times that volume of stillage [1800 m³/d (500,000 gal/d)].

Table 4.38. Material balance of typical independent sugarcane distillery (basis: 1 m^3 anhydrous alcohol)

Item	Quantity
Sugarcane, tons	15.1
Chemicals, kg	46.2
Fusel oil, kg	4.8
96° GL alcohol, liters	50.0
Stillage, tons	12.0
Bagasse, tons	3.8
Carbon dioxide, kg	760.0

Reprinted by permission of Yang, V., and S. C. Trinidade, 1978. "The Brasilian Gasohol Program," in *Energy from Biomass and Wastes, IGT Symposium*, Washington, D.C., Aug. 14–18, 1978, Institute of Gas Technology, Chicago, Ill., Table 2, p. 827.

In Brazil the waste is either lagooned or sprayed over the sugarcane fields (Yang and Trinidade 1978). Despite the absence of pathogens in stillage, it has sufficient BOD and odor to require handling similar to that used in treating municipal sewage. However, stillage clearly should be assessed for use rather than for disposal in order to improve alcohol production economics and eliminate possible environmental problems (Sect. 6.4).

Corn and grains. Producing ethanol from corn or grains rather than from sugarcane has advantages and disadvantages (Lipinsky et al. 1977*b*). Corn or grain can be stored, whereas sugarcane must be processed immediately after harvest. However, corn requires a preliminary treatment to hydrolyze or saccharify the starch into fermentable sugars. In general, saccharification involves dilute acidification or treatment with enzymes (Sect. 4.2.3.3). The ground corn kernels are mixed with water to form a mash that is cooked with steam under pressure to solubilize and gelatinize the starch (Prescott and Dunn 1949). Acid or enzymes are added after removing the mash to a converter tub and reducing the temperature to about 60°C (140°F). After conversion of the starch to sugar, fermentation proceeds just as for sugarcane mash. With corn, no bagasse is available for firing the mill boilers, so an external fuel source is needed. This need for fuel places a severe strain on the process energetics because as much energy is required to distill a gallon of alcohol as is present in a gallon of the final product. However, using coal or wood as a fuel allows the conversion of a more available fuel into a more desirable (liquid) transportation fuel. Additionally, the fermentation of corn or grains yields, as by-products, both brewer's grains (an animal feed) and brewer's yeast (a source of B vitamins used widely in the food industry for enriching white flours and products made from white flour).

Wood. Converting wood to fermentable sugars involves more severe pretreatment than does converting starch to sugar (Blake and Salo 1977). Hammer-milled wood chips are discharged into a digester into which live steam and sulfuric acid are introduced, raising the temperature and maintaining it at 149°C (300°F) for 15 min. After draining the liquid from the digester, fresh sulfuric acid is introduced, and the digester is pressurized to 1.7 MPa (250 psia) at 149°C (300°F) for 70 min. This procedure separates the lignins, furfural, and methanol. The hot sulfuric acid hydrolysate is neutralized with lime, producing a calcium sulfate slurry that is separated in a clarifier and disposed of. Fermentation of the wood sugars proceeds as for sugarcane mash.

Table 4.39 gives the characteristics of a 770 metric ton/d dry-wood fermentation plant (Blake and Salo 1977). The plant produces 250 metric tons/d of 190-proof ethanol and 94 metric tons/d of calcium sulfate sludge for disposal. The plant can generate air emissions and water effluents, depending on the fuel (coal or wood) used, to raise 238,000 kg/h (525,000 lb/h) of steam and 5000 kW of electrical power. Slops would also be produced

Table 4.39. Characterization of an 850-ODT/d wood biomass feedstock acid-hydrolysis and fermentation plant for ethanol production[a]

Characteristics	Evaluation
Wood biomass	
Feedstock	850 ODT/d
Moisture content	50%
Heat input	1.445×10^{10} Btu/d
Product output	
Before fermentation	
Hexoses	536 tons/d
Pentoses	56.7 tons/d
Furfural	16.6 tons/d
Methanol	13.8 tons/d
After fermentation	
Ethanol (190 proof)	279 tons/d
Waste	0.677×10^{10} Btu/d[b]
Residual sugars and pentose	54 tons/d
Calcium sulfate	104 tons/d
Site requirements	
Steam	525,000 lb/h
Electrical power	5,000 kW
Water (evaporation)	633,000 gal/d
Sulfuric acid 100%	92.6 tons/d
Lime (calcium carbonate)	52.5 tons/d
Diammonium phosphate	0.4 tons/d
NH_3	6.9 tons/d
Land area	15.0 acres

[a]ODT = oven dry tons.
[b]Heating value (12,780) (0.95) = 12,141 Btu/lb gross.
Source: Blake, D., and D. Salo, 1977. *Systems Description and Engineering Costs for Solar-Related Technologies*, vol. IX, *Biomass Fuels Production and Conversion Systems*, p. 98, Mitre Corp., Metrek Division, McLean, Va.

by the fermentation, but the character might differ from sugarcane or corn stillage.

4.5.4.2 Small-scale fermentation. Small on-farm stills for producing fuel alcohol to be used onsite may be encouraged as a means of establishing energy self-sufficiency on the farm. The most expedient farm feedstock for fermentation is corn; except for Carr (1972), few descriptions of small fermenters were located. Like the on-farm anaerobic digester, on-farm stills are expected to be simple, inexpensive, and easy to operate, and to have low manpower needs. Effluents are expected to be the same as for a commercial facility except on a much smaller scale and to be controlled as determined by the individual owner-farmer. However, the current legal costs and constraints associated with producing alcohol, irrespective of use, are so overwhelming that they completely discourage the activity on a small scale (Portola Institute 1974).

References

Arola, R. A., 1977. "Fuel Definition and Analysis," pp. 43–52 in *Wood Energy,* M. L. Hiser, ed., Ann Arbor Science Publications, Inc., Ann Arbor, Mich.

Beall, S. E., et al., 1974. *An Assessment of the Environmental Impact of Alternative Energy Sources,* ORNL-5024, Oak Ridge National Laboratory.

Beck, S. R. (Texas Technological University), 1979. Personal communication to H. M. Braunstein.

Biomass Energy Institute, 1978. *Biomass Energy Stories: A Portfolio Illustrating Current Economic Uses of Renewable Biomass Energy,* Report HCP/T0285-01, prepared for the U.S. Department of Energy.

Blake, D., and D. Salo, 1977. *Systems Description and Engineering Costs for Solar-Related Technologies,* vol. IX: *Biomass Fuels Production and Conversion Systems,* Mitre Corp., McLean, Va.

Braunstein, H. M., E. D. Copenhaver, and H. A. Pfuderer, 1977. *Environmental, Health, and Control Aspects of Coal Conversion,* ORNL/EIS-94, Oak Ridge National Laboratory.

Calvin, M., 1976. "Photosynthesis as a Resource for Energy and Materials," *Am. Sci.* **64:** 270–78.

Capar, S. G., et al., 1978. "Multielement Analysis of Animal Feed, Animal Wastes, and Sewage Sludge," *Environ. Sci. Technol.* **12**(7): 785–90.

Carr, J., 1972. *The Second Oldest Profession: An Informal History of Moonshining in America,* Prentice-Hall, Inc., Englewood Cliffs, N.J.

Chiranjivi, C., 1978. "Design Analysis of Small-Scale Anaerobic Digesters in India," pp. 449–64 in *Symposium on Energy from Biomass and Wastes, Washington, D.C.,* Institute of Gas Technology, Chicago, Ill.

Cioffi, P. L., 1978. *Corn/Coal Fuel Characterization Study,* Report HCP/T3052-01, U.S. Department of Energy, Washington, D.C.

Del Fosse, E. S., 1977. "Water Hyacinth Biomass Yield Potentials," pp. 73–99 in *Clean Fuels from Biomass and Waste Symposium Papers,* Institute of Gas Technology, Chicago, Ill.

DiNovo, S. T., et al., 1978. *Preliminary Environmental Assessment of Biomass Conversion to Synthetic Fuels,* EPA-600/7-78-204, Industrial Environmental Research Laboratory, Cincinnati, Ohio.

Doerr, R. C., A. E. Wasserman, and W. Fiddler, 1966. "Composition of Hickory Sawdust Smoke. Low Boiling Constituents," *J. Agric. Food Chem.* **14**(6): 662–65.

Ecklund, E. E., 1978. *Comparative Automotive Engine Operation when Fueled with Ethanol and Methanol,* Report HCP/W1737-01, U.S. Department of Energy, Washington, D.C.

Einhorn, I. N., et al., 1973. *The Physiological and Toxicological Aspects of Smoke Produced During the Combustion of Polymeric Materials,* NSF-RA-E-73-196, National Science Foundation RANN Program, Washington, D.C.

Ellis, E. L., 1962. "Inorganic Constituents of Wood," *For. Prod. J.* **12:** 271–74.

Fraser, H. S., and E. P. Swan, 1972. *Chemical Analyses of Veneer-Dryer Condensates,* Report VP-X-101, Department of Environment, Canadian Forestry Service, Western Forest Products Laboratory, Vancouver, British Columbia.

Garabedian, H. T., and C. R. Sanborn, 1978. "Environmental Aspect of Wood Fuel," pp. 19.1–19.21 in *Proc. Am. Nucl. Soc. Symp. on Environmental Aspects of Nonconventional Energy Resources II, Sept. 26–29, Denver, Colo.,* American Nuclear Society, LaGrange Park, Ill.

Garrett, D. E., 1977*a*. "Thermochemical Processing of Biomass Materials," pp. 173–208 in *Clean Fuels from Biomass and Waste Symposium,* Institute of Gas Technology, Chicago, Ill.

______, 1977*b*. *Conversion of Biomass Materials into Gaseous Products, Phase 1,* SAN/1241-77/1, Garrett Energy Research and Engineering, Inc., Claremont, Calif.

Genung, R. K., et al., 1979. "Pilot Plant Demonstration of an Anaerobic Fixed-Film Bioreactor for Wastewater Treatment," in *Biotechnology in Energy Production and Conservation,* S. C. Scott, ed., Wiley, New York.

Gikis, B. J., et al., 1978. *Preliminary Environmental Assessment of Energy Conversion Processes for Agricultural and Forest Product Residues,* EPA 600/7-78-047, NTIS PB-281 189, Municipal Environmental Research Laboratory, Cincinnati, Ohio.

Golembiewski, M., 1979. *Environmental Assessment of a Waste-to-Energy Process Wood and Oil Fired Power Boiler,* Midwest Research Institute, Kansas City, Mo.

Golueke, C. G., W. J. Oswald, and H. B. Gotaas, 1957. "Anaerobic Digestion of Algae," *Appl. Microbiol.* **5:** 47–55.

Hagen, D. L., 1976. *Methanol: Its Synthesis, Use as a Fuel, Economics and Hazards,* M.S. thesis, University of Minnesota, NP-21727, U.S. Energy Research and Development Admin., Washington, D.C.

Hall, E. H., et al., 1976. *Comparison of Fossil and Wood Fuels,* EPA-600/2-76-056, Battelle Columbus Laboratories, Columbus, Ohio, p. 55.

Hart, M. R., et al., 1976. *Ocean Food and Energy Farm Kelp Pretreatment and Sepaation Processes, Final Report to U.S. Naval Weapons Center, China Lake, Calif.,* Western Regional Research Center, Agricultural Research Service, U.S. Department of Agriculture, Berkeley, Calif.

Hillman, W. S., and D. D. Culley, Jr., 1978. "The Uses of Duckweed," *Am. Sci.* **66:** 442–51.

Hipkin, H. G., and D. J. Basuino, 1978. *Syngas from Manure—A Conceptual Plant Design,* Bechtel National, Inc., Research and Engineering, San Francisco, Calif.

Hoagland, D. R., 1915. "Organic Constituents of Pacific Coast Kelp," *J. Agric. Res.* **4:** 39–58.

Hodam, R. H., and R. O. Williams, 1978. "Small Scale Gasification of Biomass to Produce a Low Btu Gas," in *Proc. of the IGT Conference, Energy from Biomass and Wastes, Aug. 14–18, 1978, Washington, D.C.,* Institute of Gas Technology, Chicago, Ill.

J. E. Atchison Consultants, Inc., 1977. *Preliminary Evaluation of New Process for Separation of Components of Sugarcane, Sweet Sorghum, and Other Plant Stalks Relative to Its Final Contribution Toward Increased Fuel and Energy Production from These Materials,* TID-28734, U.S. Department of Energy, Washington, D.C.

Jackson, G. A., 1977. "Biological Constraints on Seaweed Culture," pp. 437–48 in *Biologi-*

cal Solar Energy Conversion, A. Mitsui et al., eds., Academic Press, New York.

Jahnsen, V. J., 1961. *The Chemical Composition of Hardwood Smoke,* Ph.D. thesis, Purdue University, Lafayette, Ind.

Johnson, A. J., and G. H. Auth, 1951. Chap. 5 in *Fuels and Combustion Handbook,* McGraw-Hill, New York.

Jones, J. L., and W. S. Fong, 1978. *Mission Analysis for the Federal Fuels from Biomass Program,* vol. V, prepared for the Division of Solar Energy Fuels from Biomass Systems Branch, U.S. Department of Energy, by S.R.I. International, Washington, D.C.

King, T. Y., 1975. *Smoke and Carbon Monoxide Formation from Materials Tested in the Smoke Density Chamber,* National Bureau of Standards, Washington, D.C.

Klass, D. L., 1977. "The Anaerobic Digestion of *Macrocystis pyrifera* under Mesophilic Conditions," in *Clean Fuels from Biomass Symposium, Orlando, Fla., Jan 25-28, 1977,* Institute of Gas Technology, Chicago, Ill.

Kohan, S. M., and P. M. Barkhorder, 1979. *Mission Analysis for the Federal Fuels from Biomass Program,* vol. IV, prepared for the Division of Distributed Solar Technology Fuels from Biomass Systems Branch, U.S. Department of Energy, by S.R.I. International, Washington, D.C.

Kreisinger, H., 1939. "Combustion of Wood-Waste Fuels," *Mech. Eng.* **61:** 115-19.

Kuo, M. C. T., and J. L. Jones, 1978. "Environmental and Energy Output Analysis for the Bioconversion of Agricultural Residues to Methane," pp. 143-76 in *Proc. of the IGT Conf. on Energy from Biomass and Wastes, Aug. 14-18, 1978, Washington, D.C.,* Institute of Gas Technology, Chicago, Ill.

Lange, H. B., Jr., 1972. "NO_x Formation in Premixed Combustion: A Kinetic Model and Experimental Data on Air Pollution and Its Control," R. W. Coughlin et al., eds., *AIChE Symp. Ser. 128* **68:** 17-27.

Lee, M. L., and R. A. Hites, 1976, "Characterization of Sulfur-Containing Polycyclic Aromatic Compounds in Carbon Blacks," *Anal. Chem.* **48**(13): 1890-93.

Lipinsky, E. S., et al., 1977*a*. *Systems Study of Fuels from Sugarcane, Sweet Sorghum, and Sugar Beets,* BMI-1957, vol. 1, prepared for the U.S. Department of Energy, Battelle Columbus Laboratories, Columbus, Ohio.

_____, 1977*b*. *Systems Study of Fuels from Sugarcane, Sweet Sorghum, Sugar Beets and Corn,* BMI-1957A, vol. 5, prepared for the U.S. Department of Energy, Battelle Columbus Laboratories, Columbus, Ohio.

Martin, G. B., and E. E. Berkau, 1972. "An Investigation of the Conversion of Various Fuel Nitrogen Compounds to Nitrogen Oxides in Oil Combustion," pp. 45-54 in "Air Pollution and Its Control," R. W. Coughlin et al., eds., *AIChE Symp. Ser. 128,* **68.**

McKetta, J. J., and W. A. Cunningham, 1978. *Encyclopedia of Chemical Processing and Design,* vol. 7, Marcel Dekker, Inc., New York.

Miller, D. L., 1976. "Methane Fermentation of Animal Wastes," supplementary comment in *Capturing the Sun Through Bioconversion,* The Bio-Energy Council, Washington, D.C., pp. 361-64.

Nelson, L. P., T. W. Lester, and J. F. Merklin, 1979. *Furnace Testing of the Combustion Characteristics and Pollutants from the Cofiring of Wheat Straw with Coal,* Kansas Engineering Experiment Station Special Report 126, Kansas State University, Manhattan, Kan.

Ott, E., H. M. Spurlin, and M. W. Grafflin, 1954. *Cellulose and Cellulose Derivatives,* Interscience, New York.

Pitt, W. W., Jr. (ORNL), June 1979. Personal communication to H. M. Braunstein.

Portola Institute, Menlo Park, Calif., 1974. *Energy Primer—Solar, Water, Wind, and Biofuels.*

Prescott, S. C., and C. G. Dunn, 1949. *Industrial Microbiology,* McGraw-Hill, New York, p. 113.

Ramos, H., S. K. Shama, and J. P. Flesch, 1972. *Health Hazard Evaluation—Toxicity Determination Report 73-7-5, Yader Locker Plant, Grantsville, Md.,* NIOSH, Cincinnati, Ohio.

Shelton, J., 1978, *The Woodburners Encyclopedia,* Vermont Crossroads Press, Waitsfield, Vt.

Skrinde, R. T., 1978. *Operation of an Anaerobic Digester at the Washington State Dairy Farm,* Tracey & Brunstiom Co., Seattle, Wash.

Thomson, W. S., 1976. "Wastewater Management in the Solid Wood and Bark Products Industries: A Review in Wood Technology, Chemical Aspects," I. S. Goldstein, *ACS*

Symp. Ser. **43**, American Chemical Society, Washington, D.C.

Thornton, T., 1978. *An Assessment of Anaerobic Digestion in U.S. Agriculture,* Report ESCS-06, Economics, Statistics, and Cooperatives Service, U.S. Department of Agriculture, Washington, D.C.

Varani, F. T., J. Burford, and R. P. Arber, 1977. "The Design of a Large-Scale Manure/Methane Facility," in *Fuels and Energy from Renewable Resources,* D. A. Tillman et al., eds., Academic Press, New York.

Wise, D. L., R. L. Wentworth, and R. G. Rispert, 1976. "Fuel Gas Production via Bioconversion," in *Capturing the Sun Through Bioconversion,* The Bio-Energy Council, Washington, D.C.

Wise, D. L., R. L. Wentworth, and E. Ashare, 1978. "Technology in Bioconversion for Producing Alternative Fuels," presented at the University of Pittsburgh School of Engineering Symposium on Alternative Resources and Technologies for Fuel Production, July 31, 1978.

Wolfhagen, J. L. (University of Maine), July 1979. Personal communication to H. M. Braunstein.

Wortman, D., 1978. "Heat Output of Wood-Burning Stoves," in *Alternative Sources of Energy 35,* Special Edition: *Wood Heat, Alternative Sources of Energy,* Milaca, Minn.

Yang, V., and S. C. Trinidade, 1978. "The Brasilian Gasohol Program," pp. 815-36 in *Energy from Biomass and Wastes, IGT Symposium, Aug. 14-18, 1978, Washington, D.C.,* Institute of Gas Technology, Chicago, Ill.

Young, H. E., and V. P. Guinn, 1966. "Chemical Elements in Complete Mature Trees of Seven Species in Maine," *Tappi* **49**(5): 190-97.

5

Potential impacts of large-scale biomass production and harvest

Production and use of biomass for energy will exploit natural resources that are highly valued for many purposes. The greatest potential negative impacts are probably those related to competition for, and overexploitation of, resources also required by the food and fiber systems, which are critical for the economic and social well-being of the United States and its international partners. The implementation of biomass energy production schemes could represent a model integration of resource management strategies for retaining and improving natural mechanisms of fertility, productivity, and regulation while minimizing detrimental side effects. Alternatively, biomass feedstock production could follow in the footsteps of agriculture and silviculture by pursuing short-term economic and energy-gain strategies, thereby creating more environmental problems.

The effects of biomass production for fuel recovery cannot yet be completely identified because of the untested status of many of the proposals. In many cases, however, the impacts of similar production processes from forestry and crop cultivation are known. The very large scale on which biomass energy systems might operate could exacerbate well-defined effects that are not now controlled or mitigated in existing production and harvest systems. The severity of increasing environmental impact would be determined by the magnitude of operations and the degree to which new and advanced mitigation techniques are built in at the start. The discussions that follow are necessarily generic. Hypothetical examples are included to help identify the nature and magnitude of potential problems.

The economic, social, and ecological issues involved in biomass production for energy are far-reaching. Careful identification of national priorities for resource use is necessary now for sound decisionmaking. Responsible application of resource management principles could make the difference between maintenance of resource quality for many purposes and degradation for all purposes.

5.1 Silviculture and Forestry

5.1.1 Effects on Terrestrial Environment

The potential terrestrial impacts of large-scale harvest and/or production of fresh biomass are considered in four categories: (1) land use, (2) soils and nutrients, (3) air quality, and (4) wildlife and ecosystems.

5.1.1.1 Land use. The large-scale harvest, or production, of woody biomass for energy could affect land use and forestry practice on millions of acres in the United States. Such changes would involve intensified management of forests; conversion of pasture, range, or cropland to forest land; and potential conflicts with agricultural production. These impacts depend on the biomass production system and regional land-use patterns. Land-use impacts are extremely important, are difficult to predict accurately, and constitute an issue that requires extensive further investigation.

Greater use of mill wastes would cause a small beneficial impact on land use through the removal of accumulated wastes. At some sites waste materials deposited over many years now occupy large areas, and the use of these wastes will allow more productive use of disposal sites.

In most cases the collection of forestry residues for energy use would cause no major change in land use; such collection would only create new patterns of forestry practice. Because of the increased demand and higher prices for timber, forest management is already being intensified by use of techniques such as site clearing and immediate reforestation; planting more productive species; site drainage; and fertilization, regular thinning, and control of fire, insects, and disease (Tillman 1978). If an increased demand for wood as fuel provided incentives for collection of logging residues, site clearing and reforestation would be facilitated. The sale of residues from thinning or logging would provide additional income and thereby enable a forest manager to make a greater investment in forest management. Thus, the use of forest residues for energy would contribute to the trend of intensified forest management.

The cutting of marginal forests would also bring about more intensive forest management. A substantial part of the privately owned forest land is held by individuals who currently have little incentive or capability to manage their land for timber production. The future is likely to hold considerable competition for wood from privately owned lands and low-grade forests. Such lands might be used for integrated harvest of timber and fuel wood through whole-tree chipping or through cutting for firewood. After cutting, such forests would undergo natural or managed regeneration or be planted as silvicultural plantations. The patterns of more intensive management that will result from increased demand for wood (both for materials and energy) have important socioeconomic implications (Chap. 7) as well as environmental consequences for soils, nutrients, and water quality.

To derive 1 EJ of wood through residue collection or cutting of marginal forests would require wood from roughly 1 to 5 million ha. These figures are based on wood densities of 12 to 54 dry metric tons per hectare (MT/ha) (Table 3.2) with the following assumptions: (1) 80% of the residues are collectable, (2) the standing biomass of marginal forests is roughly equivalent to residue from commercial forests, and (3) the energy value is 1.76×10^{10} J/MT. In very rough terms then, producing 1 EJ of wood from residues or marginal forests could cause land-use impacts (as described above) on 0.3 to 2% of all U.S. forestlands.

Land requirements for silvicultural energy farms (SEFs) were investigated in the Mitre Corporation study (Salo et al. 1977). Three suitability criteria were applied: the land must (1) be arable, (2) receive at least 63.5 cm (25 in.) of precipitation per year, and (3) have gentle to moderate slopes (<30%). Using land-use data from the Soil Conservation Services (SCS) 1967 Conservation Needs Inventory, six categories of land were defined (Table 5.1). Energy yields were calculated by assuming that 10% of the land in these categories was planted in SEFs.

The Mitre study did not investigate the availability of land for SEFs. It seems likely that SEFs would be constrained by the difficulty of either (1) procuring large tracts of land or (2) managing many discontinuous smaller tracts. Nevertheless, the Mitre study provides an overview of potential land-use impacts.

In the Mitre study, scenario 1 considers conversion of SCS class IV land to SEFs. This land includes 33 million ha (82 million acres) and is considered marginal for crop production. Scenario 2 considers all the permanent pasture, forest, and range in SCS classes I-IV, a total of 108 million ha (268 million acres). For both of these scenarios, the land taken for SEFs is mostly in the southeastern, Lake, southern Plains, Appalachian, northeastern, and Delta states regions. Scenario 3 adds fallowed and rotation cropland, while scenario 4 includes almost all cropland. Under these scenarios greater impacts would be felt in the midwest states. The wetlands scenario

Table 5.1. Total land area and projected energy yield from 10% of the area in Mitre's land availability scenarios

	Scenario[a]	Total land area (acres × 10^6)	Energy yield from 10% of land [quads (10^{15} Btu)][b]	
			Current	Future
1	SCS class IV — permanent pasture, forest, range	82	1.1	2.1
2	SCS classes I—IV — permanent pasture, forest, range	268	3.8	7.0
3	SCS classes I—IV — permanent pasture, forest, range, rotation hay/pasture, hayland, open land formerly cropped	325	4.5	8.3
4	SCS classes I—IV — permanent pasture, forest, range, all cropland except conservation use only	490	6.8	13.7
5	SCS classes II_w—IV_w — permanent pasture, forest, range	90	1.3	2.3
6A	USFS classes I—III — national forest system land	12	0.2	0.4
6B	USFS classes I—IV — national forest system land	30	0.5	0.9

[a]SCS = Soil Conservation Service; USFS = U.S. Forest Service.

[b]Yield varies by U.S. Department of Agriculture farm production region. The current weighted average yield is about 8 dry ton equivalents/acre-year, and the future weighted average yield is about 15 dry ton equivalents/acre-year.

Source: Salo, D. J., et al., 1977. Table on p. 5—7 in *Silvicultural Biomass Farms*, vol. III, *Land Suitability and Availability*, MTR-7347, Mitre Corporation, McLean, Va.

considers 36 million ha (90 million acres) limited by wetness, predominantly lands in the southeastern, Delta states, Lake states, and northeastern regions. The Pacific and northern Plains regions contribute little land to SEFs, except in the national forests system scenario, which considers national forest system lands.

The Mitre study concluded that the greatest impact of SEFs would probably occur in the Lake states, southern Plains, Delta states, and southeastern regions. Land in the Corn Belt would probably be used for intensive agriculture, whereas climate, terrain, ownership patterns, and urbanization would limit use of land for SEFs in the northeastern and Appalachian regions. Converting noncroplands to SEFs (scenarios 1 and 2) would have minimal direct impact on crop production. Despite this fact, SEFs would convert forest, pasture, and rangeland. This conversion could create land-use conflicts with agriculture, both directly and because such noncroplands are the areas most likely to be converted to cropland if necessary (see Sect. 5.2.1). Scenarios 3 and 4 would clearly involve direct land-use conflicts with agriculture.

Because the size, location, and operating mode of SEFs are somewhat uncertain, it is difficult to project what land-use impacts would result from large-scale establishment of such energy farms. However, assuming that SEFs produce 10 to 35 dry metric tons/ha-year (4 to 14 dry tons/acre-year), the land requirements to produce 1 EJ of wood would range from 1.6 to 5.7 million ha. In addition to conflicts with forestry, agriculture, and perhaps agricultural biomass-energy production, establishment of large SEFs would require acquisition or control of large land areas. Thus, changes in the patterns of land ownership or leasing could be important. The land-use impacts of SEFs clearly require further investigation, including assessments for various regions and specific projects.

5.1.1.2 Soils and nutrients. Most of the impacts of conventional forest harvesting (Sect. 2.1.5) would occur in production of wood for energy. The cutting of marginal forests or SEFs for energy would produce impacts similar to those of conventional logging. Wood production for energy would also involve increased residue removal, increased whole-tree utilization, and intensive site preparation for SEFs. These practices would produce unique impacts, in addition to the typical impacts of logging. Figure 5.1 shows general potential impacts of forest harvesting. The

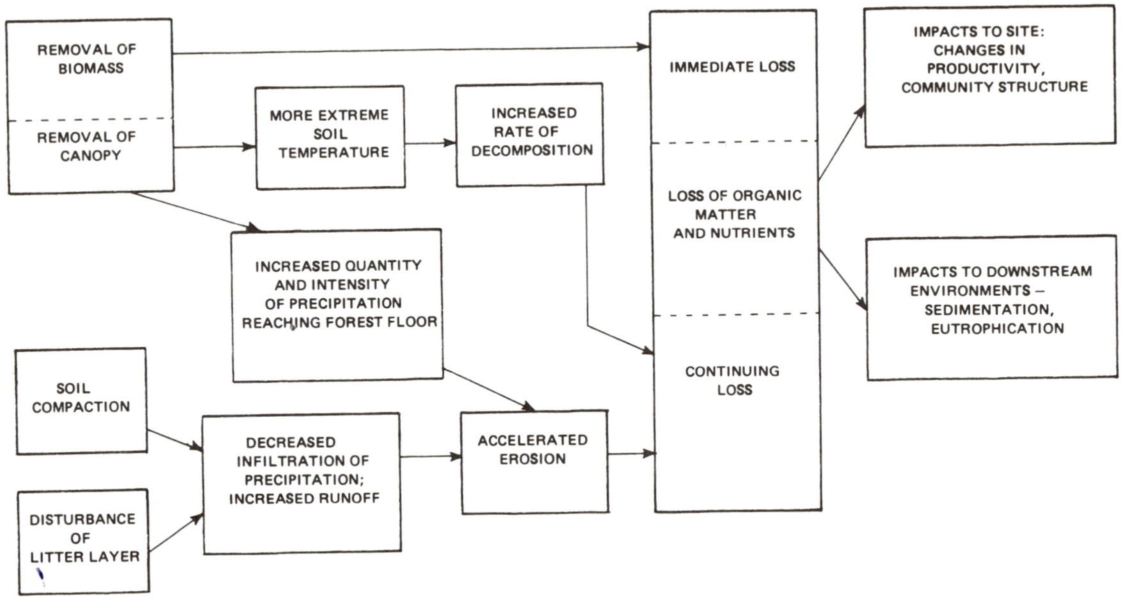

Fig. 5.1. Potential impacts of biomass harvesting on soils and nutrients.

mechanisms and consequences of these impacts are described in detail below, and the intensity of impacts is related to the type of operation.

Soil disturbance. Soil disturbance can lead to erosion, which is a major contributor to soil depletion and may adversely affect downstream environments (Sect. 5.1.2). The degree of soil disturbance can vary greatly with the type of operation and the practices employed.

Equipment for handling forest biomass runs on rubber tires or tracks (endless metal belts). Vehicles with rubber tires are slightly less costly, but tracked vehicles can negotiate slopes of up to 30% or more, although often with substantial soil disturbance. Both types of equipment can compact soils and seriously disturb soil litter. Disturbance of the soil litter layer would be greatest in skid trails and vehicle tracks. The degree of soil compaction depends on several factors, especially the weight of the vehicle. Soil compaction occurs most readily when soils are wet. In the New England and North Central states, harvesting is often accomplished when the ground is snow covered, thus greatly reducing soil compaction and litter disturbance.

Soil compaction decreases the airspace within soils, thus decreasing the moisture-holding capacity, the ability of roots to penetrate soil, and the success of seed germination and seedling survival. Both soil compaction and disturbance of the soil litter contribute to decreased infiltration of precipitation, increased runoff, and the risk of accelerated erosion (DiNovo et al. 1978).

In conventional logging, trees are felled, the branches and tops are removed, and the bole is dragged or carried to a landing or to a loading zone. However, if residue collection or whole-tree harvest is desired, the most effective method may be to fell and carry the tree to the landing for bucking and/or chipping. Ideally, this could be accomplished with the same amount of vehicular traffic as for conventional logging. However, the additional collection of undersized and cull trees could require increased traffic and cause increases in soil compaction and litter disturbance. The use of mobile hogs to pick up and chip slash and cull timber could also lead to increased soil disturbance.

In an experimental whole-tree chipping operation in Vermont (Bradley and Stevens 1978), nearly 100% of the ground surface in clear-cut areas was disturbed. Eighty-three percent of the sample plots showed major disturbance of the humus layer and exposure of the mineral soil. Disturbance was greater than would be expected for conventional logging of stemwood because of the dragging of branches and foliage across the forest floor. Areas that were selectively cut showed less general soil disturbance, but rutting and gully erosion occurred in skid trails from repeated use. This research project is continuing, and the evaluation of soil compaction, litter disturbance, and erosion rates is not yet completed.

As currently envisioned (Howlett and Gamache 1977*a*), creating SEFs would require soil disturbance during site preparation and stand establishment. Site preparation would involve some or all of the following steps: clearing; removing or reducing residue; disking or plowing to expose the mineral soil, fertilizer, and herbicide application; and planting by use of tractor-drawn machines. The objectives of these activities are to remove physical impediments to tree growth and machinery movements, to control competing vegetation, and to aerate the soil (Howlett and Gamache 1977*a*). These activities would involve greater soil disturbance than conventional logging operations. The extent of accelerated erosion from increased soil disturbance at a specific site would depend on factors such as slope of the site, the time until vegetative cover is reestablished, and—perhaps most important—the practices used in site preparation and harvest.

Nutrient removal. The immediate impact of harvesting forested systems is the removal of organic matter and nutrients contained in the harvested material. In conventional forestry, rotation times are long, logging slash is left in place so that nutrients remain in the forest, and depletion of nutrients or soil organic matter has seldom been a major problem. However, the prospect of intensified harvest of woody biomass for energy has prompted concern over soil depletion.

The harvesting of forest residues could increase the direct removal of organic matter from the forest by more than 50% over normal timber harvest. Howlett and Gamache (1977*b*) estimate that slash constitutes 25 to 43% of the weight of a merchantable bole. Collection of forest residues, however, would include cull timber, which is abundant in some forests. The removal of tree stumps would take approximately 10% additional biomass over the weight of merchantable wood. The actual increment in biomass harvested by collecting forest residues would depend on the thoroughness of collection, which would, in turn, be much influenced by the economics. Some fraction of residue will be uncollectable under any circumstance.

Schemes for residue collection and harvest of marginal forests or SEFs are likely to involve whole-tree utilization or at least taking a larger fraction of total tree biomass. The decision as to whether or not branches and foliage are harvested is an important factor in determining direct nutrient removal. Nutrients are most concentrated in the extremities of trees and least concentrated in the stemwood (Duvigneaud and Denaeyer-de Smet 1970; Fortescue and Marten 1970). Although there is much variation with different nutrients and species, nutrients are often over ten times more concentrated in leaves or branches than in the wood from the trunk. Thus, the biomass gain from leaf and branch harvesting must be weighed against the cost in nutrient removal.

Direct nutrient losses from whole-tree harvesting are compared with losses from conventional logging in Table 5.2. These data were developed by JPR Associates, Inc. (1975), for the state of Vermont, based on data from the Hubbard Brook watershed experiment. Table 5.2 compares the weight of nutrient elements in the aboveground biomass of all trees in Vermont with the stemwood and bark of growing stock trees greater than 5 in. in diameter at breast height. For most of the elements considered, whole-tree chipping would cause at least an eightfold increase in nutrient removal. These data are specific to Vermont, however, where the growing stock represents only about 65% of the total standing forest biomass. Other estimates are more conservative. Morrison and Foster (1979) reviewed stand nutrient content data for several parts of the world and indicate only a two- to fourfold increase in nutrient removal with whole-tree harvesting. Similarly, Wells and Jorgensen (1979) estimate only a two- to threefold increase in nutrient removal with whole-tree harvesting in several stands throughout the United States. It should also be noted that even whole-tree harvesting will remove less than 5 or 10% of the total nutrient capital of most forest stands (since the largest

Table 5.2. A comparison of estimated nutrient losses from conventional versus whole-tree logging

Element	Whole-tree logging	Conventional logging	Ratio (whole tree/conventional)
Ca	700,790	88,909	7.9
Mg	82,607	11,480	7.2
K	37,045	24,908	1.5
P	57,200	3,483	16.4
S	63,746	6,651	9.6
N	550,620	48,736	11.3
Na	4,295	672	6.4
Fe	6,792	716	9.5
Zn	8,310	807	10.3
Cu	623	64	9.8
Mn	87,590	11,731	7.5
Ash weight	2,743,084	316,390	8.7

Note: Figures are based on Hubbard Brook data applied to the Vermont forest inventory and represent tons of nutrients in the Vermont forest standing crop.

Source: JPR Associates, Inc., 1975. Table 1.13, p. 32, in *The Feasibility of Generating Electricity in the State of Vermont Using Wood as a Fuel: A Study*, submitted to the Vermont Agency of Environmental Conservation, Department of Forests and Parks, Stowe, Vt.

reservoir is in the soil) and that much of this can be offset by nutrient inputs via precipitation over the rotation period.

In SEFs, nutrient removal will be accelerated not only by whole-tree use but also by shorter rotation periods. Natural nutrient inputs are probably insufficient to offset losses, so fertilizer supplements are expected to be necessary, although interplanting with nitrogen-fixing species is being investigated for maintaining soil fertility.

Indirect nutrient loss. Although the harvesting of biomass directly removes nutrients and organic matter, indirect losses can occur by erosion and leaching of nutrients from disturbed soils. Litter disturbance and soil compaction increase the risk of accelerated soil erosion. Removal of vegetation can also increase erosion by increasing the amount and velocity of precipitation that reaches the forest floor.

Another cause of nutrient loss from deforestation is the leaching of elements and their export through the hydrologic system. In investigations at Hubbard Brook Experimental Forest, a clear-cut and herbicide treatment resulted in drastic increases in nitrification and nitrate-cation leaching (Likens et al. 1969). Much lower levels of nitrate leaching were observed, however, following commercial clear-cutting *without* herbicide treatment on nearby watersheds (Likens et al. 1978). Aubertin and Patric (1974), Brown et al. (1973), and Waide and Swank (1977) noted increases in nitrate leaching following forest harvesting in West Virginia, Oregon, and North Carolina, respectively. Only in the Hubbard Brook studies, however, did nitrate concentrations reach values that would contribute significantly to the total anionic component of stream waters.

If sufficient losses of organic matter and nutrients occur, the productivity of the site may ultimately decline. The amount of soil depletion that would produce loss of productivity depends on many factors, such as (1) the preexisting soil content of nutrients and organic matter, (2) the requirements of the specific tree species, (3) any fertilizer inputs, (4) natural rates of nutrient regeneration, and (5) rotation period. The risk of loss of productivity because of intensified harvest is highly site specific; assessing this risk will require analyses of soil content, standing crop of organic matter and nutrients, and knowledge of the turnover rates. Such analyses are especially important for micronutrients as well as for macronutrients. Much additional research, both general and site specific, is needed to monitor the long-term effects of intensive harvest.

5.1.1.3 Air quality. The collection and use of forestry residues would reduce air emissions from both controlled residue burning and wildfires. In the Pacific Northwest, where the greatest quantities of residues are created, burning is the principal method of residue disposal (Jemison and Lowden 1974). Broadcast burning to dispose of logging slash is being replaced by piling and burning. Overall, burning has decreased somewhat, in compliance with air quality requirements. However, the failure to burn residues is believed to increase the risk of wildfire, resulting in both uncontrolled production of smoke and damage to forests (Cramer 1974). Table 5.3 compares airborne emissions from open burning with those from furnace combustion. If collecting forest residue for energy production replaces burning, some of the air emissions from burning in forests will be replaced by emissions from stoves and boilers (see Sect. 4.4.1). To the extent that combustion of residues for energy is better controlled and produces fewer emissions than slash burning, a beneficial air-quality impact will result. Increased removal of logging slash for energy uses may also reduce wildfires and their large, uncontrolled air emissions. Determining the impacts of these trends will require analysis on a region-by-region basis.

Air emissions from internal combustion engines would be produced in collecting forestry residues, in harvesting marginal forests or SEFs, in chipping wood, and in transporting wood or chips to the point of use. Gikis et al. (1978) analyzed the emissions that would result from procuring 2712 metric tons per day (3000 tons per day) of green residues, the amount of fuel needed to fire a power plant with an approximate output of 70 MW(e). Chipping would create emissions of 2400 kg/d (5304 lb/d) of hydrocarbons, NO_x, and 4816 kg/d (10,608 lb/d) of carbon monoxide. With diesel truck transport (75 round trips) over a maximum one-way haul distance of 93 km (50 miles) and an average distance of 46 km (25 miles), daily emissions would be 78 kg of NO_x, 108 kg of carbon monoxide, 17 kg of hydrocarbons, 10 kg of sulfur dioxide, and 5 kg of particulates. Compared to emissions from combustion of the wood chips (Table 4.22), the particulate and NO_x emissions from chipping and transport are insignificant. If we assume midrange values for both CO and HC combustion emissions (15 kg/metric ton of material burned, or 30 lb/ton), then handling will produce roughly 10% of the CO and 5% of the HC produced by combustion. These emissions

Table 5.3. Comparison of emissions from open burning of forest residues with wood/bark boiler emissions

(pounds per ton burned)

	Open burning[a]	Wood and bark in boilers[b]
Particulates	11—58[c]	30
Carbon monoxide	40—140	2—60
Hydrocarbons	4—18	2—70
Nitrogen oxides	2—4	10
Sulfur oxides	—	1.5

[a]Data compiled by Cramer, O. P., 1974. "Air Quality, Fire-Hazard, and Esthetics," pp. F1—F51, in *Environmental Effects of Forest Residues Management in the Pacific Northwest — A State-of-Knowledge Compendium*, O. P. Cramer ed., USDA Forest Service Gen. Tech. Report PNW-24, Pacific Northwest Forest and Range Exp. Sta., Portland, Ore.

[b]Data from U.S. Environmental Protection Agency. *1973 Compilation of Air Pollutant Emission Factors*, AP-42, 2d ed. (and supplements), Research Triangle Park, N.C.

[c]The 58 lb is for wildfire.

would be dispersed over a wide, predominantly rural area, and the impacts would depend on the local air quality, topography, and other factors.

Finally, the establishment of SEFs would involve site preparation, soil disturbance, and possible production of fugitive dust. These air emissions would principally depend on the length of time that mineral soils would be exposed. Other important factors would be the size of the tract, local wind patterns, and any mitigation measures undertaken.

5.1.1.4 Impacts on wildlife and forest ecosystems. The large-scale use of forest biomass for energy would bring about changes in forest habitat and wildlife (Fig. 5.2). Habitat alteration can produce long-lasting impacts to wildlife, and if large areas of habitat are affected, changes in the types, distribution, and density of wildlife can be profound. Because some wildlife species are migratory, the impacts of habitat alteration are sometimes surprisingly far reaching.

The principal habitat changes associated with collection of forestry residues for energy are the removal of slash, cull trees, and snags (standing dead and partially dead trees). Slash removal can have either adverse or beneficial impact, depending on the species affected. Certain wildlife species, especially woodpeckers, feed on insects from under the bark of slash, and slash may provide cover to ground-nesting species and their predators. However, for most birds of eastern hardwood forests, logging slash appears to be relatively unimportant as a site for cover, nesting, or food production (Kroodsma and Shugart, 1975).

Several rabbit species benefit from the presence of slash, and increases in their populations can have significant secondary benefits to other important wildlife species. For example, cottontail rabbit populations can be maintained or increased by leaving slash on the ground or by forming brush piles from the slash (Anderson and Pelton 1976, Haugen 1943). Denser cottontail populations provide

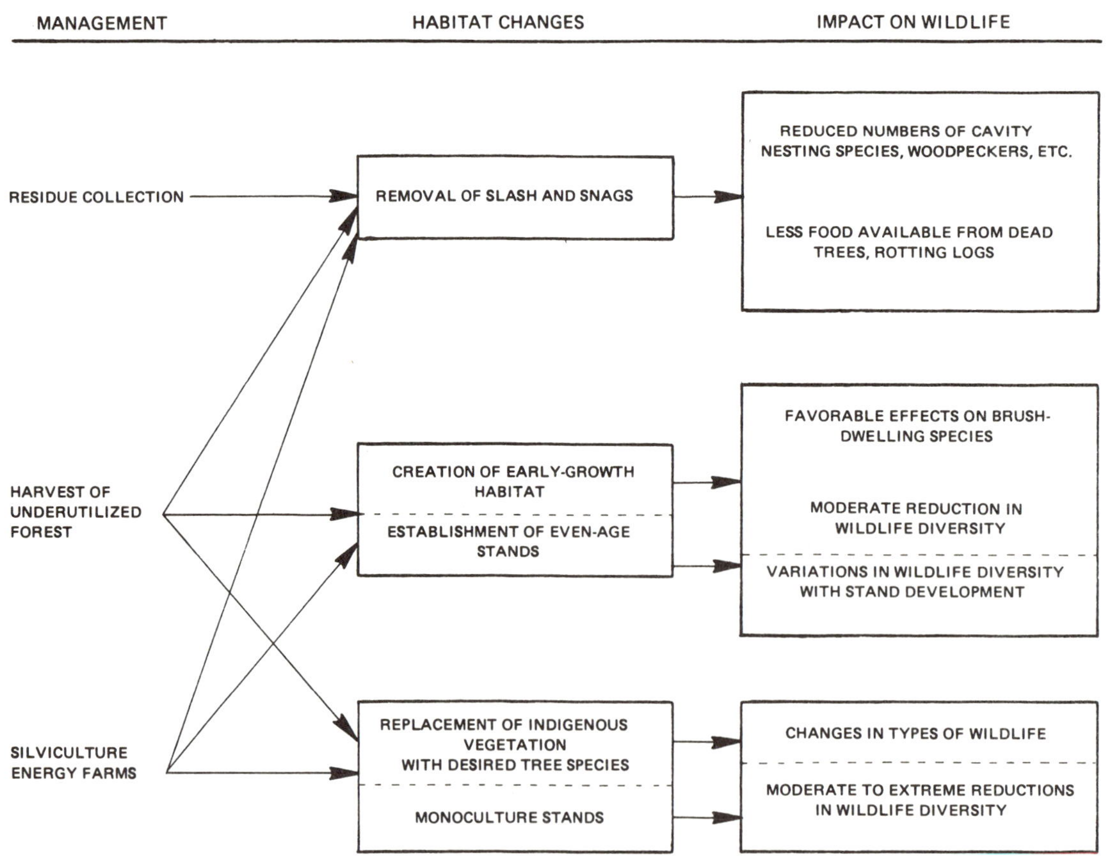

Fig. 5.2. Potential impacts to wildlife of large-scale biomass energy harvest and/or production.

greater food production for a large number of important avian and mammalian predators, including foxes, bobcats, and several species of hawks and owls.

The importance of snags and large cull trees is widely acknowledged, and the required number and distribution of snags for several bird species have been quantified (Evans and Conner 1979). Raphael and White (1978) found that snags were used by over 30% of the birds and mammals on the western slope of the Sierra Nevada. Snags served as sites for nesting, roosting, denning, foraging, and perching. In Missouri, the snag-dependent species (cavity nesters) make up 20% of the bird species encountered on the national breeding bird survey (Evans and Dawson 1976). In the northeastern United States, more than 25 bird species nest in cavities and depend on snags (Evans and Conner 1979). Thus the collection of snags for energy would reduce habitat for wildlife. Reduction of habitat for cavity nesting species may have indirect effects. Many of these species are insectivorous and may help control insect populations that damage timber. Evans and Conner (1979) concluded that "economic advantages of providing areas with mature forest habitat for cavity nesters may at first seem remote. However, evidence suggests that long-term advantages may exist ... and many studies have demonstrated (cavity nesters') effectiveness in reducing populations of insect pests that attack and destroy trees ..."

If forests are clear-cut for their energy content, natural or managed revegetation will follow or a silvicultural plantation may be established. Revegetation would create more xeric brush habitat, characteristic of early successional stages. This would cause the loss of wildlife species preferring forest interiors but would favor brush-dwelling species. Opening the forest canopy promotes the growth of ground cover and increases the availability of some types of food for species such as deer, rabbits, and quail. However, large clear-cuts may reduce suitable habitat for deer, because of this species' tendency to remain near the forest edge (Segelquist and Rogers 1975, Blymyer and Mosby 1977) and because of the loss of annual mast crops. Turkeys would be adversely affected, because they prefer extensive mature open forests (USDA Forest Service 1971) and open grassy fields such as pastures (Hamrick and Davis 1972). The removal of treetops eliminates habitat for species such as squirrels and cavity-nesting birds. Because brush-dwelling bird species are only about 20% of the total number of breeding species in the eastern mountains, Kroodsma and Shugart predict that clear-cutting large portions of Appalachian forest would decrease overall bird diversity. However, by creating many small clear-cuts within larger areas of medium-aged and mature forests, wildlife diversity in the eastern forest could probably be increased.

The creation of even-age stands would cause fluctuations to occur in wildlife density as the stands developed. The patterns of wildlife fluctuation vary with the forest type, but the general pattern for hardwood forest succession is as follows. Wildlife density in a clear-cut area would be low for a few years after harvest. As shrubs and brush developed and food became available, wildlife density would increase but would decline again in the pole-stage forest. Wildlife diversity would rise again as the forest matured, but many highly managed stands would be harvested before this stage. Through the entire process, the species composition would be continually changing for all types of wildlife.

The impacts to wildlife of residue collection or whole-tree harvesting can be mitigated to some extent by leaving some snags standing, especially large ones. In stands that have had snags removed, trees could be left and/or killed for snags (Raphael and White 1978). Diversity of birds can be raised by limiting the size and distribution of clear-cuts and the fraction of forest area in early growth stages.

An SEF would involve major modifications of natural habitat. Management with short rotation periods would maintain vegetation predominantly in brushy stages. Competing plant species would be controlled, thus limiting the types of vegetation and species diversity. Although experiments with mixed-species plantings could be conducted, initial SEFs would probably be planted in a single tree species, perhaps even a highly selected variety or clone within a species. The habitat created and maintained in this way would be structurally and perhaps genetically highly simplified. In contrast with the mature forest, the simplified habitat of a silvicultural plantation does not support a wide variety of wildlife (Leopold 1978). Furthermore, the large areas of land currently considered marginal for forestry or agriculture are very important as wildlife habitat. If such areas were converted to plantations managed solely for energy crops, severe adverse impacts on wildlife could result (Benemann 1978). On the other hand, it may be possible to design and operate SEFs to accommodate the habitat requirements of some wildlife species, which would probably require some sacrifice in the rate of biomass production but could result in beneficial impacts to some wildlife. Additional research is needed to determine the ultimate impacts of SEFs on wildlife and the optimal site-specific design and operation for SEFs.

The large-scale production of wood for energy would involve some major ecological uncertainties, such as the stability of simplified production systems and the possible effects on the global carbon cycle or climate (Sect. 6.5). Although it is not possible to project specific environmental impacts associated with these issues, these questions need to be addressed by further research and monitoring. If large areas are brought under moderate to intense management for the single purpose of maximizing biomass harvest, it is likely that the resulting systems will be less stable than natural forests or forest ecosystems managed for multiple use. Silvicultural energy farms that do not provide sufficient genetic heterogeneity may be susceptible to outbreaks of pests or disease organisms. Maintenance of intensely managed production systems, especially monocultures, requires substantial inputs of energy, fertilizers, biocides, and labor (Cox and Atkins 1975). The problems of ecological stability in managed systems are comparable to the ecological problems of agriculture, as discussed in Sect. 5.2.1.4.

5.1.2 Effects on Aquatic Environment

Large-scale collection or production of biomass for energy would involve impacts on (1) hydrology and water use, (2) water quality, and (3) aquatic biota. In general, the use of mill wastes should not affect aquatic environments.

5.1.2.1 Hydrology and water use. The principal hydrologic impact of collecting or producing woody biomass for energy would be increased runoff from lands receiving intensified management. Residue removal, whole-tree chipping, and establishment of silvicultural plantations would each produce distinct patterns of alteration. Silvicultural energy farms might also affect water use through irrigation.

When rain enters a forest, it is first intercepted by leaves and branches, and then it evaporates or infiltrates the soil. Very little overland runoff occurs, except from roads, compacted soils, and other disturbed areas. Water enters streams primarily through the soil. Forest cutting reduces both evaporation and transpiration (the passage of water from soils through plants into the atmosphere) (Stone 1973). Thus, the cutting of trees increases the amount of water moving into soils, reduces the amount of water removed by plants from soils, and increases the amount of water moving from soils into streams. The hydrologic yield from a watershed increases after cutting in proportion to the reduction in forest canopy. Stream flows return to normal as forest regeneration occurs. Forest cutting apparently has only minor effects on the timing and magnitude of flood peaks, provided that good logging practices are used and that cutting does not affect a large proportion of a large watershed.

The hydrologic impact of residue removal would be a slight increase in the watershed's yield over what would have occurred with logging alone. Slash removal would result in (1) reduction in surface area for evaporation and (2) possible increased soil compaction from increased use of collection equipment. If silvicultural plantations or marginal forests are clear-cut as an energy source, there will be moderate increases in water yield. In a study at the Fenrow Experimental Forest in Parsons, West Virginia, water yield increased from 17 to 25% in the first year after commercial clear-cut logging (Kockenderfer and Aubertin 1975).

Silvicultural energy farms may be irrigated to ensure the success of stand establishment and to promote maximum rates of growth during dry periods when trees are under water stress. Irrigation might use either flooding systems or sprinkler systems, depending on terrain, soil type, and costs of the system. The amount of irrigation water delivered would vary with the regional and seasonal weather conditions, the availability of water, and the operating costs. Impacts of irrigation might include competition with agriculture for water supply, reduced stream flow from consumptive withdrawals, and drawdown of groundwater levels.

5.1.2.2 Water quality. In the harvest or production of forest biomass for energy, water quality could be affected by sedimentation; logging debris in stream channels; changes in stream temperatures; and release of nutrients, organic matter, or silvicultural chemicals. These threats to water quality are classified as "nonpoint sources," and impacts are heavily dependent on the forestry practices employed. Under the U.S. Environmental Protection Agency's implementation of Section 208 of the Clean Water Act (PL 92-500), best management practices to control pollution for silvicultural activities are being developed.

Sedimentation of streams occurs as a result of erosion. As noted in Sect. 5.1.1.2, erosion is primarily associated not with tree-cutting but with logging roads and skid trails. In areas where mineral soils are exposed or compacted, sheet erosion can detach and transport soil particles offsite. Sediment-bearing runoff often releases its sediment load as it flows through an undisturbed area. Thus, this type of erosion often has little effect on stream water quality if proper buffer zones are left along stream banks (Stone 1973). The extent of sedimentation impacts depends greatly on the design of road systems and skid-trail networks and on the care used in logging. Studies of clear-cutting in a West Virginia forest indicate that, where logging was conducted carefully, stream turbidities rarely exceeded 5 ppm, whereas in a clear-cut watershed that was carelessly logged, stream turbidity averaged 490 ppm and reached a peak of 56,000 ppm during one rain storm (Reinhart et al. 1963).

Stream temperatures can increase dramatically if shade-producing vegetation is removed from stream banks. For instance, when stream-bank vegetation was removed during clear-cutting of a Douglas-fir watershed in Oregon, monthly mean stream temperatures were 7 to 12°F above normal between April and August (Levno and Rothacher 1967). Such temperature increases can be avoided for the most part by leaving a strip of shading vegetation along streams in clear-cut areas.

Streams are also susceptible to adverse impacts from physical alternations such as the disturbances caused by bridge and road construction, cross-stream skidding, and placement or accumulation of debris in stream channels. These disturbances can cause sedimentation and degradation of habitat for aquatic organisms. Planning and good practice can prevent or minimize these impacts to a large extent.

The removal of logging residues for energy would probably cause little stream sedimentation over that which would occur with conventional logging operations. If poor logging practices produce substantially accelerated erosion, the additional disturbance created by residue collection might contribute significantly to increased sedimentation. For the most part, however, residue collection would cause only small to moderate intensification of erosion risk. Removal of slash would also limit the impact of debris accumulation in streambeds.

The water quality impacts of opening unharvested forest to energy harvest would depend heavily on whether logging roads were built. Roads create the largest portion of the erosion problem, perhaps as great as 90% in many cases (Megahan 1972). Building roads is often a major cost in logging operations, especially well-planned road systems that are designed for long life and minimum impacts (Stone 1973). Because the economic return from harvesting marginal forests for energy is likely to be lower than the return from logging, the lack of roads may preclude harvesting. In other cases landowners may be willing to invest in construction of access roads, based on the prospects of economic returns from activities beyond the initial harvest. The construction of new roads and the possible use of old roads in poor condition create the possibility of significant increases in erosion and stream sedimentation.

Much of the unharvested forest land in the United States is privately owned and receives little or no management. Because of this, one cannot assume that the best management practices would be uniformly applied in harvesting this land. The opening of inaccessible marginal forests for energy harvest could cause 10- to 100-fold increases in erosion from the affected area, with subsequent impact on streams. Good planning and adequate control measures could restrict this impact greatly. Where planning and good practice are lacking, however, the rate of erosion could cause temporary sediment yields approaching those from agricultural cultivation.

Silvicultural energy farms would incur the impacts of site preparation and stand establishment in addition to the impacts of harvesting. The removal or reduction of residues and the scarification of soils might allow short-term runoff, erosion, and sedimentation. If revegetation were slow, the impacts of site preparation might extend over several years. The uses of mechanical cultivation or herbicides to control vegetation that competes with the crop species might retard the tendency for erosion rates to return to normal levels. The rates of erosion and sedimentation for SEFs would greatly depend on the erodibility of the soils and the management practices employed. Quantification of the long-term impacts of SEFs will require monitoring and research on a regional and site-specific basis.

The intensive management of SEFs might also affect water quality through the use of fertilizers, herbicides, and insecticides. Use of these silvicultural chemicals involves the difficult task of balancing the benefits, costs, and risks of damage to humans, wildlife, and aquatic ecosystems.

Fertilizers would probably be applied to SEFs, both to enhance plant productivity and to maintain soil fertility over the long term. Fertilizer requirements will vary greatly with tree species and soils of a given site, and research is continuing into the most effective fertilization system (Howlett and Gamache 1977*a*). In conventional silviculture, fertilization occurs largely in the Pacific Northwest, where nitrogen deficiencies are often encountered, and in the Southeast, where lack of phosphorus often limits tree growth (Oregon State University 1977). Fertilizers are usually applied from aircraft, and this pattern would probably prevail for SEFs of large area. Urea and phosphate are the forms of fertilizer commonly applied (Oregon State University 1977). Although urea is soluble, it is rapidly converted to ammonia and eventually further oxidized to nitrate. Both ammonia and phosphates are fairly rapidly absorbed by soils or taken up by living plants or microorganisms. The pattern for appearance of nutrients in surface waters varies greatly with the season and weather, but generally there is a large peak in the concentration of phosphate, urea, or perhaps ammonia within a few days after application. This peak is caused by the direct input of fertilizer to water surfaces and any immediate solubilization, leaching, and/or runoff. Further loss may cause slightly elevated nitrate concentrations for several weeks or months. A lesser secondary peak of nitrate may appear in stream water following a period when soils are saturated (Oregon State University 1977).

In a study by Fredriksen et al. (1975), a 68-ha watershed in southwestern Oregon was fertilized in March 1970 with 224 kg of urea-N per hectare. Maximum concentrations of 1.39 ppm urea-N and 0.04 ppm NH^4-N were reached 48 h after application started. Nitrate-N concentration reached 0.17 ppm after three days and returned to pretreatment levels within two months. A second peak of nitrate-N occurred seven months later with fall rains, and elevated concentrations persisted for several months.

Herbicide use in SEFs would vary with the crop species and weed species. Over 200 chemicals are currently classed as herbicides, and these, used singly and in combinations, vary in their selectivity, mode of application, and persistence in the environment. Preemergence herbicides are applied to bare soil; contact herbicides are applied to foliage or other living surfaces. Some herbicides are injected into tree trunks, but this mode of application is not likely to have an impact on water quality (Oregon State University 1977). Some of the most commonly used herbicides are 2,4-D, Amitrole, Picloram, and, prior to restrictions on its use, 2,4,5-T. In some cases, herbicides are applied by aerial spraying; for example, 2,4-D and 2,4,5-T are applied from aircraft to promote stand establishment of Douglas fir. In hardwood plantations, herbicides must often be applied by tractor or by hand, and spray must be directed at weeds to avoid damages to foliage or buds of the crop species (Howlett and Gamache 1977*a*). The mode of application greatly influences the distribution of herbicides. Herbicides can enter streams in drift or through direct application of spray materials to surface waters, in overland flow, in surface-water runoff, or in subsurface water flow through the soil (Fredriksen et al. 1975). The magnitude and duration of stream contamination is markedly different for each of these processes (Table 5.4). For instance, Picloram is sufficiently soluble and persistent to reach stream water from subsurface flow. The two most widely used herbicides, 2,4-D and 2,4,5-T, however, are relatively immobile in soils and are nonpersistent. In many cases stream contamination after application of herbicides is brief, limited to several days or weeks.

Table 5.4. Magnitude and duration of herbicide entry into forest streams

Mechanism of entry	Magnitude of peak residues (mg/liter)	Duration of entry
Drift or direct application to stream surface	>0.01	hours
Overland flow	<0.01	days
Leaching and subsurface flow	<0.001	weeks

Reprinted by permission of Fredricksen, R. L., D. G. Moore, and L. A. Norris, 1975. "Impact of Timber Harvest, Fertilization and Herbicide Treatment on Stream-Water Quality in Western Oregon and Washington," Table 9, p. 306, in *Forest Soils and Land Management, Proc. of the 4th North American Forest Soils Conference, Laval University, Quebec*, B. Bernier and C. H. Winget, eds. Les Presses de l'Université Laval, Quebec, Canada.

Howlett and Gamache (1977*a*) suggested that integrated pest management would be used to protect SEFs from insect damage. Strategies would include site preparation, fertilization, and water management, tailored to give tree species maximum resistance to pests. Insecticides would be used to control or prevent serious pest outbreaks. A wide variety of chemicals are available for this purpose. The choice of an insecticide, its formulation, and the mode of its application would depend on its toxicity for the insect in question, insecticidal selectivity, hazard to man and other nontarget organisms, persistence, cost, speed of action, and other factors. Insecticides could be applied aerially or from the ground, in the form of powders, liquids, or aerosols. As with herbicides, application from aircraft provides the greatest opportunity for insecticides to enter surface waters. Such application typically produces a pulse contamination event with many factors influencing the size and the duration of the pulse.

Clearly, the use of fertilizers, herbicides, and pesticides in SEFs would involve many possible impacts on water quality, including impacts on aquatic organisms (Sect. 5.1.2.3) and on the use of water for irrigation, public supply, or other purposes. The impacts of using silvicultural chemicals depend greatly on site-specific circumstances such as the chemical applied, local topography, the season of application, and weather conditions at the time of application. Impacts will also depend on management practices such as whether strips of vegetation are left along streams or whether operators avoid direct application of chemicals to stream channels. Predicting impacts of using silvicultural chemicals will require extensive continuing analyses for various regions, sites, and projects.

5.1.2.3 Fish and aquatic ecosystems. With harvest or production of woody biomass for energy, impacts to aquatic ecosystems would result from water quality changes such as (1) sedimentation, (2) changes in water temperature, (3) enrichment of nutrients and organic matter, and (4) toxic concentrations of silvicultural chemicals. In practice, stream ecosystems are usually simultaneously affected by two or more of these perturbations. Sediments can affect fish by clogging their gills and covering spawning beds. Increased water temperatures can eliminate fish such as trout, which require cold-water streams. Additions of nutrients can increase primary production and change the species of algae present, whereas the addition of organic matter associated with sediments can cause oxygen depletion through increased microbial metabolism. The stresses induced by all these factors can increase the susceptibility of fish to disease or to the toxic effects of pesticides or herbicides.

Impacts to aquatic ecosystems will probably occur in proportion to water quality alterations (Sect. 5.1.2.2). Because of its relatively minor impact on water quality, the removal of forest residues would probably produce very little impact on aquatic ecosystems. Most adverse impacts would be associated with logging or with reforestation efforts, and collection of residues would probably intensify sedimentation to only a minor extent. The cutting of forests that would otherwise have remained standing could cause small to moderate adverse impacts from sedimentation in large numbers of forest streams. However, the greatest potential for adverse impacts would result from site preparation and practices used to establish and maintain SEFs. If planning and control measures are inadequate, the combined effects of sedimentation, fertilization, and chemical applications could seriously degrade water quality and the quality of downstream aquatic ecosystems.

5.2 Agricultural Crops

5.2.1 Impacts on Terrestrial Environment

5.2.1.1 Land use. The conversion of large amounts of resources, particularly land, from current uses to production of biomass for energy feedstocks is a potential source of conflict between the energy producers and the American agricultural system. For example, to produce an ethanol supply sufficient to displace 10% of the gasoline now used annually without diversion of grain presently committed to export or livestock consumption would require production of additional fermentation feedstocks. Corn is a logical candidate for these feedstocks since it is our most efficient crop (Sect. 2.2.3) and can be cultivated over a large geographic area. To supply the required additional 110 million metric tons, corn cultivation would have to expand by about 20 million hectares. If, as is likely, the land in question is of poorer quality than present cropland, this figure could be considerably higher since per-hectare yields would be lower. Twenty million hectares represents almost one-half the amount of land now in the cropland reserve (USDA 1977) (Sect. 2.2.5.1).

As a second example, the theoretical "energy plantation" discussed in Sect. 3.2.2.2 (Henry, Fraser, and Vail 1976) was suggested to have the capacity to produce an estimated 10 EJ (or 14% of the nation's 1974 energy consumption) per year by cultivation of perennial grasses. At a realistic estimate of yield for improved (fertilized) pasture, such a plantation would require about 150 million hectares of pastureland to produce a sufficient quantity of grass feedstocks. This is about 40% of the land now used for livestock grazing in the United States.

The point to be made from these examples is that sunlight is diffuse and photosynthetic energy storage is not very efficient. Therefore, the collection of even a moderate amount of energy from biomass could require a huge commitment of land. The ultimate risk, that of competing for the use of land suitable for food and fiber production, could be the same whether or not the land in question is currently used for crop production. This is the case in light of two major factors. First, agricultural land is already being lost at a rate that has prompted substantial concern on the part of federal and state agencies (Council on Environmental Quality 1978; Carter 1977; U.S. Environmental Protection Agency 1978; see Sect. 2.5.1). This land may have to be replaced by conversions from the cropland reserve to maintain present agricultural production levels. Second, the cost of fossil fuels is rising steadily as supplies dwindle, and it is probable that the agricultural system will be unable to continue to rely on massive energy subsidies to increase productivity, although demand for agricultural commodities will almost surely continue to grow in the future. Furthermore, greatly elevated fuel costs or absolute shortages could force our agricultural system to reduce the amount of energy it already uses by modifying intensive practices. Less-intensive agriculture will require more extensive use of land if total production is to be maintained; in other words, there will be a trade-off of land for energy. This trade-off may also bring food and fiber production into serious, long-term conflict with production of biomass for energy, even though biomass crops are not now envisioned as occupying land that is *currently* producing food. This situation is especially the case because the economics of biomass energy production requires maximal efficiency of energy input and because land that is most appropriate for biomass cultivation (Sect. 3.2.2.2) is also likely to be the land to which food and fiber production would expand if needed.

Similarly, conversion of pasture or rangeland to biomass production could compete with the raising of livestock. Range-fed meat production is more than ten times as energy efficient as confinement production. If energy and cost constraints in the future take their toll on agricultural output, greater reliance on rangeland for meat production will result. Unconfined meat production is more efficient precisely because of the large amount of rangeland currently available (Green 1978). Should this availability substantially decrease, conflict over the land resource could arise.

5.2.1.2 Soils and nutrients. When returned to the soil, crop residue and the soil humus resulting from its decomposition (1) retain plant nutrients, (2) keep soil porous and in condition for easy tillage and good plant growth, (3) increase infiltration of water into the soil and hence increase water storage, (4) reduce soil detachment caused by wind and raindrop impact, and (5) reduce velocity of runoff and associated soil transport potential (Larson 1979). Crop residue is thus an important tool for the maintenance of soil quality and control of wind and water erosion (Larson et al. 1978). The use of crop residue for fuel recovery will compete directly with soil conditioning uses under many conditions. The impacts of residue removal could vary considerably with location, ranging from negligible to severe, depending on site or regional conditions.

Perhaps the most extensive constraints on removal of crop residues stem from the need to protect soil from erosion. Impacts of erosion are difficult to quantify because they are influenced by factors such as crop variety, soil nutrients, soil structure, topsoil depth, drainage, temperature, moisture, and pests. Evidence suggests, however, that each 1-cm loss of topsoil (or a loss of about 150 metric tons/ha) from a basis of 30 cm (12 in.) represents a decrease in productivity that is conservatively estimated at 250 kg/ha of corn, 86 kg/ha of oats, 110 kg/ha of wheat, and 175 kg/ha of soybeans (Pimentel et al. 1976). With soil loss levels commonly 20 to 40 metric tons (or 0.13 to 0.27 cm) per year, yield reductions may be considerable given cumulative effects over many years. Soil is reformed at much lower rates than typical loss rates. Under ideal soil management conditions, 1 cm of soil could reform in about 12 years; but under normal agricultural conditions, 1 cm reforms in about 40 years.

Protection against erosion is considered adequate when soil removal by wind or water is less than or equal to soil-loss tolerance limits (Larson 1979). These limits are defined by SCS as the maximum allowable soil loss that is safe for maintaining long-term soil productivity. Maximum soil-loss tolerance ranges from 2 to 10 metric tons per hectare (1 to 5 tons per acre) per year, depending on soil properties, soil depth, topography, and prior erosion (Gupta et al. 1976). Use of the universal soil loss equation or the wind erosion equation has allowed estimates of permissible residue removal levels for some regions. The need for water erosion protection would limit residue removal to 58% of the total residue produced in the Corn Belt (Lindstrom et al. 1979); 40% in Georgia, the Carolinas, and Virginia; and 10% in Mississippi and Alabama (Campbell et al. 1979). Wind erosion protection only would leave 21% of produced residues available for removal from the Great Plains. If, however, water erosion is considered with wind erosion, available residues from this region are even lower (Skidmore et al. 1979).

Removal of residues may also conflict sharply with maintenance requirements for soil nutrients. Residues from nine leading crops (barley, corn, cotton, oats, rice, rye, sorghum, soybeans, and wheat) are estimated to contain about 4×10^6 metric tons of nitrogen, 0.5×10^6 metric tons of phosphorus, and 4×10^6 metric tons of potassium (Larson et al. 1978). This amount is the portion of nutrients extracted from the soil by crop vegetation that can be recycled after harvest (Holt 1979). These amounts are roughly equivalent to 40% of the nitrogen, 10% of the phosphorus, and 80% of the potassium currently applied in inorganic fertilizers (Larson et al. 1978). Hence, residues represent an appreciable portion of the nutrients applied to cropland. Removal of these nutrients would necessitate their replacement via energy-expensive inorganic fertilizer. The annual gross fertilizer value of nitrogen, phosphorus, and potassium in total available crop residues is estimated at $3.3 billion (USDA 1979). The energy required to manufacture additional fertilizer could substantially decrease the net energy gain obtainable from residues as fuel. Furthermore, if removal of residue resulted in accelerated erosion, even larger amounts of nutrients would be lost indirectly because eroded surface soil contains more nutrients by weight and is thus more fertile than the underlying material from which it is derived (Holt 1979). Hence, nutrient depletion could be both the direct and indirect result of removing residue for fuel recovery. Depending on the conversion processes employed, however, use of residues for fuel recovery need not render them valueless as fertilizer; for example, anaerobic digestion would preserve substantial portions of nutrients for reapplication to the land. Pyrolysis or direct combustion would, however, result in some nutrient loss.

The use of crop residue for management may have other important effects. Residue on soil surface increases organic-matter content and thereby reduces runoff from storms of all sizes and eliminates runoff from small-storm precipitation (Onstad and Otterby 1979). Increases in the amount of storm water retained in soil become increasingly important along a gradient of light (sandy) soils to heavy (clay) soils, which have the highest runoff potential. Clay soils in the Corn Belt, for example, will retain an extra 25 mm (1 in.) of water annually, which could mean an extra 630 kg of corn per hectare (10 bu/acre) with proper crop residue management (Onstad and Otterby 1979). In fact, water storage capacity is found to be the soil profile characteristic that contributes most to improved crop yields in systems of residue handling (Van Doren and Allmaras 1978). The high organic-matter content provided by residues also improves tilth and can reduce plant pathogen populations by stimulating growth of competitive or antagonistic soil microorganisms (USDA 1979).

Conversely, residues remaining on the soil surface can decrease the productivity of some soils under certain climatic conditions. In climatic regions where residues retard soil warming in the spring, residues should not cover the planted row (Van Doren and Allmaras 1978). Residues can also release toxic compounds as they undergo microbial decomposition and thereby reduce the yields of some crops (particularly wheat) under wetter climatic conditions (Elliott et al. 1978). Excessive amounts of residues from some crops (such as wheat, corn, and sugarcane) can accumulate because microbial decomposition is suppressed by increased loading rates under some soil and climatic conditions (Parr and Papendick 1978). Some residues (e.g., rice) are burned because they interfere with tillage and planting. Such burning can result in air quality problems (Sects. 5.1.1.3 and 4.4.1). Thus, the intensive concentration of residues in some regions might provide part of the energy needs of those regions without detriment to agricultural productivity.

Preliminary studies indicate that the amount of crop residue that can be safely removed is extremely site specific. This amount is determined partly by uncontrollable climatic, soil, and topographic factors for a given area and partly by choice of cropping and management practices that can be varied to maximize conservation of soil (Larson et al. 1978). Conservation-tillage or no-tillage schemes can increase safe levels of residue removal. Combining minimum-tillage practices with the practice of allowing some residue to remain may be particularly effective. For example, the estimated percentage of cropland adequately protected against water erosion in the Corn Belt increases from 36% under conventional tillage (plow, disc, harrow) with all residue removed to 78% for no tillage with 3920 kg of residue left per hectare (Lindstrom et al. 1979). Only site-specific studies will be able to determine all the necessary information to safeguard against loss of productivity.

The use of residues for fuel recovery may also compete with their use as feed for livestock. The U.S. Department of Agriculture (USDA) estimates that 3.4 metric tons of feed for each of the 116 million head of beef and dairy cattle in the United States could be derived from crop residue (USDA 1979). Competition between these uses is not likely to become serious, however, as long as conventional forage is available at low cost and residues become increasingly valuable in conservation and no-tillage systems.

Figure 5.3 summarizes the potential effects on soils and nutrients of residue removal or whole-plant harvesting techniques. Site-specific conditions will determine the severity of the impacts depicted. Current research is directed toward understanding the critical interactions between residue removal rates

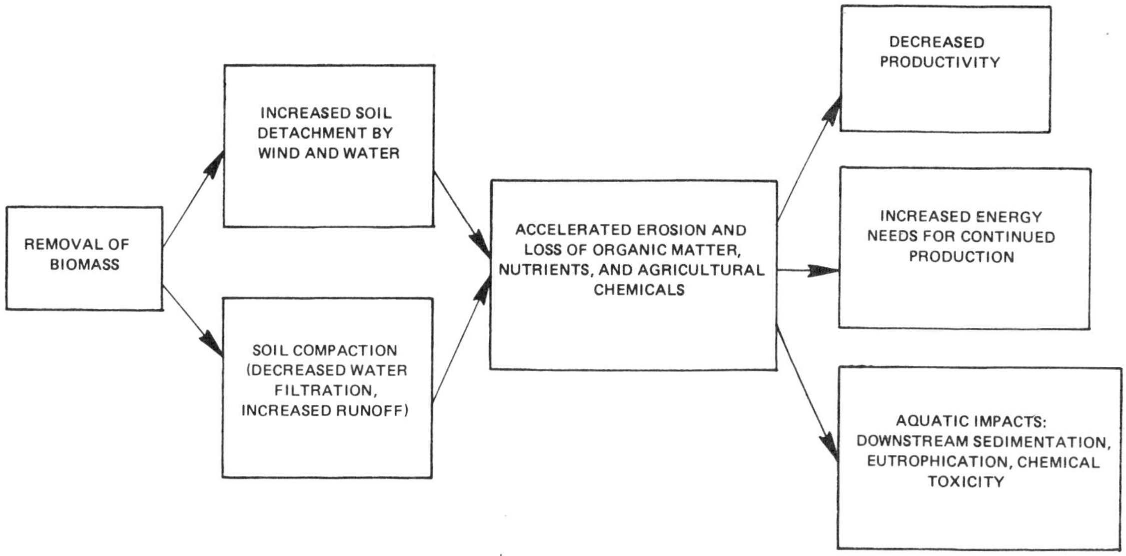

Fig. 5.3. Potential impacts of agricultural residue removal or whole-plant harvesting on soils and nutrients.

and soil fertility so that safe levels of biomass harvest may be predicted on a site-specific basis.

The use of agricultural residue for energy production has the advantage of not requiring commitment of additional land. Production of new biomass crops, however, is likely to require use of land deemed marginal for agriculture (see Sect. 3.2.2.2). When cultivated, marginal lands are generally more susceptible to erosion than are present croplands. The SCS currently recognizes 2 million ha (5 million acres) of land that should be retired from cropping (USDA 1977). Thus the impetus to bring more marginal land into cultivation contradicts USDA policy. This might not be true, however, if biomass production were to concentrate on crops most ecologically appropriate for the region in which they were raised (e.g., sod-forming grasses in the Great Plains) (Lockeretz 1978) and if farming practices that minimize soil disturbance (e.g., no-till farming) were employed. Although crops that are most ecologically appropriate may not be the most productive in terms of fuel recovery maximization, the overriding concern should be for land quality and correct matching of crops to land type so that soil degradation is minimized (Doering 1977). In the long run, this is thought to be the only feasible approach, economically and ecologically. Doering (1977) points out that it makes little sense over the long term to switch from pasture to continuous row crops under conventional tillage only to induce erosion that ultimately degrades land quality for both uses.

5.2.1.3 Air quality. Large-scale production of biomass for fuel recovery could involve the application of agricultural practices to extensive new areas of land. This expanded cultivation would intensify the kinds of air quality impacts that are familiar from standard agricultural production (e.g., emissions of fugitive dust from tilling and airborne pesticides). In addition, use of marginal lands may entail dust problems due to wind erosion that are especially hard to overcome. For example, wind erosion in the Great Plains has been a major source of air pollution in the form of recurrent severe dust storms in the 1930s, 1950s, and 1970s (Lockeretz 1978). Average dust concentrations were found to be 10 to 15 mg/m^3 during dust storms in the driest areas of the southern Great Plains, while average dust concentrations ranged from about 5 to 10 mg/m^3 even in the least erosive areas. These concentrations are very large compared with the 1.0-mg/m^3 concentrations considered severe in urban air pollution episodes (Hagen and Woodruff 1973). Atmospheric particulate mass from wind erosion in the United States probably already exceeds the combined output from all other primary sources which emit 35.2 million tons of particles annually (Hagen and Woodruff 1975). The magnitude of potential increase in suspended particulate pollutants with increased use of marginal land cannot be accurately estimated at the present time. Data from the Great Plains during the period of "all-out production" of wheat in the mid-1970s suggests, however, that there is a direct relationship between marginal acreage planted and acreage damaged by wind erosion. If this relationship is true for other regions, a proportional increase in general atmospheric particulate pollutant load may be expected.

Air quality impacts are mitigated by the same techniques that are applied to reduce soil erosion by wind. Such techniques include planting wind breaks and leaving crop residue on the soil to minimize the soil's exposure. In addition, employing management schemes designed to minimize energy inputs could also be beneficial in mitigating air quality impacts. "No-till" farming is a management scheme that leaves soil undisturbed for several seasons and harvests only aboveground plant parts. Unfortunately, this technique also requires the use of greater quantities of pesticide. The result could be that control of particulate pollution will result in greater pesticide pollution of both air and water.

Removal of crop residue for fuel recovery could also be largely antagonistic to mitigation of both soil erosion and air pollution. Site-specific factors must be examined to determine safe levels of residue removal for a given location. Removing crop residues that typically are burned in open fields would, however, alleviate the air quality impacts associated with this uncontrolled source of emissions.

5.2.1.4 Wildlife and ecosystems. Agriculture has perhaps produced the most profound impact on wildlife of all human activities in the United States. Agriculture has affected huge areas over a brief time span, a combination of factors that has allowed wildlife little or no time to adapt and no room for retreat (Burger 1978). An excellent review by Burger for the Council on Environmental Quality traces the history of agricultural impacts on wildlife. Clearing forest and grassland habitats of natural vegetation and replacing it with farms and homesteads long ago resulted in reduction, or even local and regional extirpation, of much of the native American fauna. By 1890, wilderness animals had already vanished in most of the East, much of the great central prairie, and the fertile valleys of the Far West. Still, the small family farm, with its fence rows, woodlots, and varied crops, maintained a certain beneficial habitat diversity and even increased habitat availability for some species, such as the bobwhite, meadowlark, red fox, woodchuck, and meadow mouse, as farmland replaced forest on a wide scale. The dramatic increase in mechanization of agriculture since World War II, however, has reversed even these trends. Habitat availability and diversity decreased to their lowest level as huge monocultures replaced the varied crops and landscapes of small family farms in order to maximize efficient operation of large specialized equipment. Pesticides have become a key ingredient in modern farming and have taken their toll on many wildlife species. The detrimental impacts of farming have been felt by all major animal groups. In some cases, the effects of agricultural practices alone have been so severe as to endanger, and even exterminate, species of invertebrates, fish, reptiles, birds, and other animals (Burger 1978).

Conversion of large new areas of land to cultivation of biomass for fuel could be a source of additional further significant change in terrestrial habitat and biota. The severity of impact will vary with the site and will probably range from minimal in areas already cleared for farming to large in habitats previously undisturbed. Highly fertile areas will be the most attractive for biomass production needs. Because it is unlikely that cropland of highest quality will be available, interest will tend to focus on natural areas with high biological productivity. The coastal wetlands of the Southeast are a good example of an environment likely to be severely affected by large-scale biomass production. Both the value and the problems of such areas are becoming increasingly well understood. Their use in the biomass effort has been proposed (see Sect. 3.2.2.1) not only because they are highly prolific but because some of these areas are among the only suitable locations for expanding sugarcane cultivation for ethanol production for climatic reasons. These factors justify a somewhat detailed discussion of this particular type of environment.

In their natural state, wetlands provide natural flood control, process organic pollutants and thereby improve water quality, and recharge aquifers. Because of their great biological productivity, wetlands also support a rich and diverse fauna, serving as nurseries for commercial fish and shellfish and as habitat for terrestrial vertebrates of commercial, recreational, aesthetic, and legal importance.

In general, the area of wetlands throughout the lower 48 states is on the decline (Council on Environmental Quality 1978). The Fish and Wildlife Service estimates that, of an initial 51 million ha (127 million acres), only 35 million ha (87 million acres) remained in 1954, and by 1971 another 7 million ha (17 million acres) had been lost (Council on Environmental Quality 1978). This decline represents an overall loss of 45%. In Louisiana alone, 1300 km^2 of coastal marsh were lost during the 30-year period prior to 1973 (Gagliano and Van Beek 1973). Dredging for coastal waterways and mineral extraction; confinement of rivers for navigation and flood control; and conversion, by filling, to urban and industrial land have all contributed to the loss. So drastic have these effects been on the Louisiana coast that it has been said that "the great Louisiana marshes are dying" (Gagliano and Van Beek 1973).

Filling and draining of wetlands may also eventually waste the land. Conversion to agriculture has been traditional, with 24% of all nonfederal agricultural soils originally wetland and 50% of nonfederal wet soils at present in the prime farmland class (Council on Environmental Quality 1978). Converted wetland soils can be highly productive, but some are totally unsuitable for cultivation once natural vegetation and drainage patterns are disrupted. For example, winter vegetable production in southern Florida is supported by drained wetland soils. The rate of soil loss by oxidation in this area approximates 2.5 cm (1 in.) per year, and these soils are expected to disappear completely by the year 2000 (Council on Environmental Quality 1978). The natural process of restoration to renew the soil could require centuries. Agricultural use has converted a once prolific natural area to a nonresource. Thus the problem with draining coastal wetlands for sugarcane production is not merely one of competition with other valuable uses but rather one of potential loss of a resource for any use at all and of acceleration of detrimental processes already well under way from other causes.

Wetlands are not the only ecosystem that could be adversely affected by biomass production. Desert lands, which could be used for cultivation of hydrocarbon crops, are another example of an environment easily injured and slow to recover. Between deserts and wetlands lies a spectrum of land types and associated ecosystems that require prudent and careful management to avert the total loss of productivity as a future resource.

The way in which biomass crops replace natural vegetation will also vary the effects on terrestrial biota. Huge monocultural stands of vegetation represent a drastically simplified biological environment. Such an environment is likely to be unattractive to most animal species because few species are adapted to a single type of cover and a single type of food. This is especially true if the vegetation in question is not native to a location. As a result, most wildlife would avoid such areas. Those animals that could survive in such a simplified environment would be apt to flourish unchecked and become pests. A monoculture would present an enormous supply of resources, would likely harbor few or no predators, and would pose no obstacles to continuous successful pest dispersal. Extensive areas of high-quality habitat have been well identified as the highest risk areas for initiation of pest outbreaks (Clark, Jones, and Holling 1978). Such outbreaks may, in turn, require massive application of pesticides that further simplify, and even sterilize, both target and nontarget environments. In short, a monocultural plantation can, in the long run, become an ecological disaster that can only be maintained by continual inputs of expensive energy. This situation could be severely detrimental to biomass production economics.

A more balanced approach would require that numerous types of vegetation, preferably native vegetation to which wildlife is already adapted, be integrated in the biomass plantation. A system of patches of different types of vegetation would provide a diversity of resources for more different kinds of beneficial animals (such as predators of pest species), would help to control the spread of pests by providing barriers to dispersal, and might also help to limit the effects of intraspecific competition among the crop plants themselves.

Agricultural production has, in one respect, been beneficial to wildlife by providing new food sources (Burger 1978). Mechanical harvest of corn, for example, leaves large quantities of waste grain in the fields. This food source has proved so attractive to waterfowl that major shifts in wintering grounds have occurred. Similarly, development of rice monocultures has provided a food source that has shifted wintering grounds from coastal marshes and river bottoms to these fields. In addition, stubble and other residue left after harvesting traditionally provide food and cover for wildlife on corn, soybean, and other crops. Residue removal and whole-plant harvest could thus have a detrimental effect on wildlife that have adapted to use of these agriculture-related resources.

5.2.2 Impacts on Aquatic Environment

5.2.2.1 Water quality and aquatic ecology. Crop production for biomass energy could have impacts on water quality and aquatic ecosystems similar to those of food and fiber agricultural production. Moreover, residue collection could intensify the impacts of agriculture itself through increased runoff from present cropland of sediments and nutrients (see Sect. 5.2.1.2). The magnitude of this increase would, of course, be highly site specific and to some extent there would be also qualitative differences from site to site. Runoff from different areas would not always contain the same constituents or the same concentrations of common constituents because of variations in soil type, tillage practices, level of pesticide application, rainfall, crops produced, etc. At a generic level, however, four major categories of water quality degradation can be identified that would result in impacts to biological communities: (1) sediment loading and siltation, (2) biological oxygen demand (BOD) loading, (3) nutrient additions, and (4) additions of pesticide and other toxicants.

Sediment loading and siltation can lead to light attenuation, thereby reducing photosynthesis and producing changes in both aquatic plant and animal communities, especially through food chain repercussions (Hynes 1971). Spawning and feeding grounds can be smothered (Reid 1961), and turbidity can reduce the ability of fish to sight food (Hart and Fuller 1974). Particulate matter may become entrapped in gills, interfering with respiration and other physiological processes. Siltation also contributes to accelerated eutrophication by fill-in of static waters (Carta and Bondurant 1976).

Annual erosion losses from United States soils now average about 20 metric tons per hectare (Larson 1979). The better, less-erodible, land is already in agricultural production and additional land brought into use for biomass cultivation would be no less erodible than land now farmed. We may expect, therefore, that soil loss will occur on at least the same average level if conventional agricultural practices are employed on new land. As an example of potential increase in water quality impact, the 20 million hectares required for a 10% gasohol program from corn (Sect. 5.2.1.1) could result in an additional 400 million metric tons of sediment eroded annually, an approximately 20% increase in stream sedimentation. Crop residue removal could also increase erosion losses. Research is in progress which is expected to provide a basis for estimating how much residue may be removed without detriment to soils and receiving waters. Such estimates at this time would be premature.

Depletion of oxygen caused by the loading of water courses with high-BOD materials can reduce species diversity by eliminating fish and invertebrates sensitive to low oxygen levels (Hynes 1971; Hart and Fuller 1974). Again, community changes occur, especially through food chain influences. Addition of chemical fertilizer and soil-borne nutrients accelerates eutrophication (Larcher 1975), leading to algal blooms (Fogg 1975), fish kills (Reid 1961), and other changes. High levels of nitrogen in the form of ammonia are also toxic to aquatic life (Hynes 1971). Again, it can only be assumed that an increase in acreage under conventional agricultural practices would result in a proportional increase in BOD loading and nutrient additions in receiving waters. The effect of residue removal on these parameters is, at this time, unknown.

Effects of pesticides and other anthropogenic toxicants are highly dependent on type, dose, duration of exposure, and organisms exposed. Pesticides may be labile or extremely stable, the latter undergoing bioaccumulation, which may produce toxicity in animals high in food chains (Hynes 1971). Pesticide application at conventional levels on a larger acreage would be expected to increase pesticide pollution. On the other hand, no-till or minimum-tillage systems reduce soil erosion but require increased use of fertilizer and pesticides. The dangers of an increase in negative water quality impacts will require that the location, the crops, and the timing and application of agricultural practices and chemicals be determined with great care for biomass energy production schemes.

5.2.2.2 Water use. Use of irrigation in large-scale biomass production could represent a significant source of competition for water resources. This competition would occur especially in regions where water is already scarce, particularly in the western half of the United States. In addition, irrigation return flows typically contain heavy loads of nutrients and pesticides that could threaten the quality of groundwater. Finally, because irrigation is extremely costly and energy intensive (Pimentel et al. 1975), biomass production that requires irrigation would probably provide little net energy gain. It is therefore unlikely that irrigation will be feasible as a practice in most large-scale biomass production schemes. This fact will tend to restrict the choice of locations to those with sufficient natural precipitation and the choice of crops to those not dependent on artificial water supplies.

5.2.2.3 Summary. In conclusion, production and use of crop biomass for fuel recovery could involve serious, long-term competition for agricultural resources and large-scale intensification of problems already associated with food and fiber production. Mitigation of impacts on soils, water quality, air quality, and wildlife may be possible by careful design of management schemes. Novel management practices may greatly improve existing technologies but will require research before application can become widespread.

Careful delineation of national priorities is needed to ensure that the resources available for biomass for both energy production and food and fiber production are partitioned in accord with the social and economic needs of *all* interest groups. Most importantly, attention to the maintenance of long-term productivity of agricultural resources is essential if biomass is to be regarded as a source of truly renewable fuels.

5.3 Livestock Wastes

Harvesting confinement-produced livestock manures for fuel recovery will have little, if any, adverse impact on the environment. Its collection and use could, indeed, have a positive effect on the environment as well as on the energy picture since much of the manure accumulated in livestock feeding operations is now regarded as waste and poses a potential environmental hazard.

5.3.1 Impacts on Terrestrial Environment

By itself, the collection of animal wastes for energy production will entail neither an increase in the amount of land devoted to feedlot operations nor a change in land-use practices. The current trend toward larger and more numerous commercial feedlots (Loehr 1974) could be reinforced by economic changes related to energy self-sufficiency and by possible new income from the sale of natural gas. It is extremely unlikely, however, that production of gas alone would cause large-scale industrial expansion.

Competition for use of manure for other purposes (e.g., for fertilizer) is more apparent than real under current circumstances. An estimated 20% of the confinement-produced manures would be sufficient to fertilize about one-third of the corn acreage in the United States (Pimentel et al. 1973), representing a potentially substantial savings in energy devoted to inorganic fertilizer production. However, payment of collection, hauling, and spreading costs would likely offset the savings.

Thus, use of manure for fertilizer would be no more economically feasible than using it for fuel production. In addition, generation of methane by anaerobic digestion of manure would not preclude its value as a fertilizer or feed supplement because usable nitrogen, protein, and other substances remain in the treated sludge. This is not true of other conversion technologies, such as pyrolysis. Digestion is therefore viewed by some agriculturalists as the *only* viable process for fuel recovery from manure (Varani and Buford 1977). Use of the sludge for other purposes would also be beneficial in waste disposal, and the costs of sludge distribution might be partially offset by the added value of the raw material as a digester feedstock. Adequately digested sludge materials do not have an offensive odor and in this way are superior to untreated manure, especially when disposal is by land spreading. Flies and other pests are also less attracted to digested wastes, and pathogenic organisms are destroyed in the digestion process (Morris, Jewell, and Casler 1975).

5.3.2 Impacts on Aquatic Environment

Water pollution from animal feedlots is related to such factors as the number of animals in the confinement unit, waste production per animal, frequency of cleaning, and characteristics of the wastes. The degree to which adjacent water bodies are affected is a function of ambient temperature, magnitude of rainfall, slope and drainage of the area, type and area of lot surface, and types of management practices (Loehr 1974). Feedlot runoff, which is rich in nitrates, ammonia, bacteria, chemicals fed to cattle (such as hormones and antibiotics), and other pollutants, may be 10 to 100 times more concentrated than raw domestic sewage (U.S. Environmental Protection Agency 1972). Under certain circumstances, feedlot runoff may affect groundwater as well as surface water by percolation, especially where aquifers are shallow and lie beneath sandy soils.

The use of manures for anaerobic digestion would not necessarily alleviate these water pollution problems. In so-called "environmental" feedlots, animals may be kept in covered facilities in which manure immediately passes through floor gratings to a collection site beneath the animal housing area. Highly efficient collection of this sort would certainly facilitate the use of manure for digestion, but digestion need not depend on any special collection method. In many older feedlots, manure will still accumulate in a widespread outdoor layer and thus may be subject to runoff before periodic collection.

Anaerobic digestion also does not eliminate the problem of wastewater disposal. A volume of wastewater comparable to that of the raw material will remain (Loehr 1974). This liquid effluent is still high in BOD, is not acceptable for discharge to surface waters, and will require treatment, storage, and handling (Thornton 1978). In any case, the problems encountered in such treatment and handling, together with the impacts of waste disposal, should be no worse than those under current management.

In summary, collection of manure for methane production is one of the most promising of the proposed uses of biomass for energy, especially from an environmental point of view.

5.4 Freshwater Systems

5.4.1 Effects on Terrestrial Environment

The schemes for large-scale production of aquatic biomass would require building new aquatic systems such as ponds and lagoons. Modification of existing water bodies is unlikely in view of the well-established multiple use of most lakes, reservoirs, and other water bodies. The conversion of terrestrial environments to aquatic systems would be the major impact of large-scale production of aquatic biomass for energy.

At present, proven technology and economic incentive do not exist for producing energy from newly grown water hyacinth. Large-scale production of energy from microalgae would depend on technological refinement, substantial reductions in the cost of harvesting, favorable rates of solar insolation, low cost of land and construction, and other factors. Energy production from freshwater biomass is not envisioned for near-term application, except perhaps when coupled with waste treatment systems. Thus, only general impacts can be projected. Nevertheless, the potential impacts on land use can be addressed.

Large-scale production of energy from either microalgae or vascular plants would require large land areas for construction of shallow ponds. To produce 6 million m^3/d (200 million scfd) of methane (an output comparable to that projected for a commercial-scale coal gasification facility), 80,130 ha (198,000 acres) of canals would be needed to grow water hyacinth (Lecuyer and Marten 1976) and 147,716 ha (365,000 acres) of algae ponds would be required (Benemann et al. 1977). Emersed vascular plants can use atmospheric CO_2 and thus have projected yields three times greater than microalgae. Although less land area appears to be needed to produce a given quantity of methane from emersed aquatic plants than from microalgae, yields from intensive culture of vascular plants or algae on a sustained basis remain to be demonstrated.

The large-scale construction of systems for producing fresh-water biomass would require relatively flat land in the southern United States. The high insolation rates required for efficient algal production suggest the use of land in the Southwest. However, because emersed vascular plants can cause very high rates of evapotranspiration, water limitations would discourage the siting of production systems in arid areas. Algal production systems would probably occur in modules 2.6 to 26 km^2 in area (1 to 10 sq miles) (CSO International, Inc. 1978). Likewise, the difficulties of acquiring large tracts of land for water hyacinth production would probably dictate smaller-size systems than the 801 km^2 envisioned by Lecuyer and Marten (1976).

Several types of land might be considered for conversion to aquatic biomass production systems. Cropland might be removed from production, but grazing land or unused land would be more readily available. Pasture, forested areas, or wetlands might be used. Land that might be considered for production of aquatic biomass includes 107 million ha (265 million acres) of cropland having excess water sufficient to limit regular cultivation (Sect. 2.4.1). Much of this land is located in the Gulf Coast and southeastern coastal area. The impacts of converting terrestrial environments to aquatic systems would depend on the type of land taken and the size of the tracts. In each case existing habitat would be destroyed by pond construction; terrestrial wildlife might be adversely affected, although the new ponds might prove attractive to waterfowl. Many small ponds would probably be less disruptive to wildlife than a few large systems because small ponds would produce an environment that is more heterogeneous and has more "edge" areas. Ponds occupying quite large areas might affect local hydrology or weather. Complete consideration of such impacts would require site-specific evaluations of proposed systems and local environments.

Although large-scale production systems for aquatic biomass would require large land areas, Oswald and Benemann (1977) suggest that they could be beneficial in soil reclamation. Digester residues could be used as soil conditioners, or "the ponds could be rotated to provide rich new fields created by deposition of algal sludge." The siting of aquatic biomass production systems on lands in need of reclamation could have beneficial impacts, and the idea needs to be further evaluated.

Schemes to use existing growth of waterweeds would probably have only minor impacts on terrestrial environments. Digesters for vascular plants would probably first be associated with wastewater treatment facilities that use aquatic plants (see Sect. 3.4). Waterweed control operations might then harvest existing growth and bring the biomass to such a facility for digestion. Land-based equipment for chopping, dewatering, storing, or loading the biomass into digesters could handle both the newly produced biomass from waste treatment and the biomass harvested from natural waters. The ponds of a typical waste treatment system could occupy a large land area, whereas the land-based handling facilities would require only a few hectares. Handling would cause impacts perhaps comparable to those of a small food-processing facility. Residuals from such a bioconversion facility are considered in Sect. 6.3.

5.4.2 Effects on Aquatic Environment

Aquatic biomass production systems could affect aquatic environments through alterations of (1) hydrologic cycles, (2) nutrient cycles, and (3) water quality. These impacts are considered for systems producing biomass solely for energy and for integrated systems that might produce clean water, feed, or food, as well as energy.

5.4.2.1 Effects on hydrologic cycles. Some hydrologic impacts will probably occur if large areas are devoted to aquatic biomass production. Water supply, sewage, or makeup water might be moved some distance to production ponds by pump or gravity flow. Interbasin transfers are conceivable. Evaporative losses might range from no loss (such as in southwestern Florida, where rainfall exceeds gross evaporation) to losses as high as 127 cm/year (50 in./year) in the desert Southwest (Dynatech 1978). In the most extreme case, as in a region with evaporation rates of 127 cm/year, 260 km^2 of ponds (100 sq miles) could theoretically lose 330 billion liters of water per year to evaporation, about 500 times the flow of the Colorado River at the United States-Mexico border. Actual evaporation rates might be somewhat less than this because the air near the large pond would become partially saturated (Dynatech 1978). Nevertheless, a large consumptive use of water could occur, even in areas with less-extreme evaporation rates. Thus, supplying large amounts of water to production ponds might require reservoir construction, water diversion projects, and/or alterations of groundwater or surface flow regimes. High evaporative losses might also affect local weather and vegetation.

Water might also be lost from production ponds by infiltration to groundwater. The need for inexpensive construction techniques might preclude use of pond liners or sealants. Experience with sewage oxidation ponds, however, indicates that self-sealing often occurs after several months of operation. The extent and impacts of leaching losses would depend on local soil and groundwater conditions.

5.4.2.2 Effects on nutrient cycles. The nutrient requirements for growing microalgae or vascular plants could be met by (1) using commercial fertilizer on a "once-through" basis, (2) using municipal sewage or similar wastewater, (3) achieving biological nitrogen fixation in the ponds and supplying phosphorus and other nutrients through other methods, or (4) recycling nutrients following the conversion process (e.g., returning anaerobic digester residues to growth pond). Nutrient recycling, of course, depends on the details of the conversion technology (Chap. 4).

A study of the costs of aquatic biomass systems by Dynatech (1978) concluded that the costs would probably make commercial fertilizer infeasible. The sewage from a city of five million people could supply sufficient nitrogen and phosphorus for a 260-km^2 (100-sq-mile) aquatic farm of water hyacinth, but cities of such size are seldom located near large areas of low-cost land. It seems more likely that sewage would be used in smaller integrated aquatic systems designed primarily for sewage treatment with a methane by-product. Such systems could achieve wastewater treatment and nutrient removal with considerable reduction of the energy normally required for conventional or advanced treatment. Because nutrients in sewage effluent often stimulate nuisance growth of plants and/or algae in waterways, a scheme that accomplishes nutrient removal in an energy-conserving manner would achieve two environmentally desirable goals at once. Of course, the overall impact of such a waste treatment-bioconversion scheme would depend on the disposition of residues from anaerobic digestion (Sect. 6.3). The ultimate fate of sewage-borne pathogens and heavy metals clearly needs to be considered.

Nitrogen could be supplied to aquatic biomass production through the activity of nitrogen-fixing blue-green algae. This process might place a penalty on productivity, either because of the lower yields accompanying algal species control (Benemann et al. 1977) or simply because of the inherently lower yields of nitrogen-fixing species, which must devote a portion of their production to nitrogen fixation (Dynatech 1978). Although large-scale nitrogen fixation in ponds for biomass production has yet to be demonstrated, it does not appear difficult to achieve. The use of algal nitrogen fixation instead of commercial nitrogen fertilizer would result in environmentally beneficial energy conservation.

Because of the high costs and the difficulties associated with the above nutrient supply systems, the incentive is great to design integrated systems for the production and conversion of aquatic biomass that return to the growth ponds any nutrients remaining after conversion. In anaerobic digestion, nitrogen, phosphorus, and other minerals are conserved almost completely. The recycle of supernatant and/or sludge from digesters to growth ponds has been tried successfully for microalgal systems (Benemann et al. 1978). Theoretical considerations (Dynatech 1978) suggest the potential feasibility of simple reintroduction (without aeration) of mixed digester residue into production ponds for vascular plants, although this area requires further investigation. A closed nutrient budget would probably be difficult to achieve. For example, small losses of nitrogen to the atmosphere through ammonia volatilization may occur. Bleed streams would probably be necessary to prevent the buildup of heavy metals or other toxicants in the system; for example, digester residue might need to be removed periodically from the production/conversion system (see Chap. 6). Some liquid effluent discharge is also possible. Nonetheless, the small amounts of nutrients lost could be supplemented to the system from the sources described above. The achievement of systems with high recycling rates would reduce environmental impacts from both discharge of nutrients and demand for energy-intensive commercial fertilizer.

5.4.2.3 Effects on water quality. Although some aquatic biomass production systems might be designed to maximize evaporation or infiltration to achieve "zero discharge," many systems are likely to discharge some aqueous effluent to a waterway. Such an effluent stream would contain nutrients, suspended solids, oxygen-demanding substances, heavy metals, or other dissolved solids. Because of the tendency for aquatic plants to take up dissolved materials and promote oxidation of organic materials, effluents from growth ponds would probably be relatively innocuous. An exception to this could occur in ponds where high evaporation rates cause accumulation of inorganic salts, as might be expected in the Southwest. In general, the composition of any wastewater and its impacts on the receiving body of water would depend on the design of the facility, the system(s) supplying the nutrients, and the local environment.

In cases where biomass harvesting is applied in sewage treatment systems, water quality may be improved. Algal sewage oxidation ponds that do not employ harvesting are typically unable to meet the Environmental Protection Agency (EPA) 1985 performance standards for secondary treatment (10 mg per liter of BOD and 10 mg per liter of suspended solids). A reliable algal harvesting technique could enhance nutrient removal and produce effluents meeting EPA requirements.

Mechanical harvesting of naturally occurring waterweeds for energy uses could also have a beneficial effect on water quality. This would result from the removal of some nutrients incorporated into the biomass and the replacement of herbicide applications with mechanical harvest.

5.5 Marine Systems

Our ability to predict potential impacts from large-scale ocean kelp production and harvesting is constrained by two factors. First, the technology associated with open-ocean kelp farming for biomass is in its infancy. The open-ocean kelp farm described in Sect. 3.5.2 is primarily the product of theoretical studies and will undoubtedly undergo major modifications before deployment. We must therefore be concerned with the generic features of the proposed system in estimating potential impacts. Second, and perhaps more restricting to this analysis, our estimate of the final size and location of kelp production systems is vague, and this is a crucial restriction as the qualitative and quantitative impacts associated with these facilities will be quite size dependent. For this analysis it is therefore essential that an assumption of size be made. The smallest single farm that appears economically feasible would cover 40,470 ha (100,000 acres) (Wilcox et al. 1976). A system has also been envisioned that would provide a substantial amount of our energy needs and would cover 5.7×10^5 km^2 (about 230,000 sq miles, or an area greater than the combined areas of South Carolina, Pennsylvania, Ohio, New York, and Florida) (Wilcox 1976). Although this latter prospectus may only be a phantasma, it will be assumed that eventual kelp production systems will be large enough to invoke the possibility of impacts associated with large systems. Although somewhat recursively defined, this assumption allows a discussion of the largest set of potential impacts. Should eventual farm deployment be of significantly smaller proportions, size-dependent impacts can be selectively scaled or eliminated from consideration.

This section treats impacts on the physical, chemical, and biological components of the environment. Many, if not all, potential impacts overlap one or more categories. Impacts are only generically and qualitatively discussed. Quantitative estimates will require site- and facility-specific information and the results of extensive modeling and analysis, which are currently unavailable. The following is, therefore, an examination of potentialities; associated probabilities have not been evaluated.

It is recognized that considerable controversy exists about the economic feasibility and potential environmental impacts associated with large kelp farms (Dynatech 1978; Jones 1979). This controversy results from the lack of baseline data, the newness of the technology, and the potential size of marine biomass farms. This analysis will be inclusive. Necessary data to separate chimera from justifiable concerns do not exist.

5.5.1 Physical Impacts

The ocean kelp biomass farms currently proposed are large and complex. Table 3.7 (Chap. 3) summarizes the material requirements for a 40,470-ha kelp farm. It will be necessary to concentrate a large support structure (more than 4 million buoys, 250,000 km of grid/support line, 5000 upwelling pipes, and more than 5000 diesel engines) in a relatively small area (about 20 km on a side). Upwelling requirements will be impressive: about 13,000 m^3/s (about 450,000 cfs), which is equivalent to the cooling water requirements for more than 200 open-cycle 1000-MW(e) nuclear plants or about 75% of the average flow of the Mississippi River. The kelp plants will be spaced at intervals of about 1 m and will trail at the surface. The 40,470-ha farm would support 400 million kelp plants. These are the requirements for one farm. To provide a significant contribution to the nation's energy supply, multiple farm deployment will be necessary. Structures of this magnitude could seriously affect the environment.

Natural stands of seaweed and kelp are known to attenuate certain frequencies of wave energy (Wayne 1976, Bascom 1964). The kelp, combined with the numerous buoys (especially those specifically designed to extract incident wave energy for the wave-powered upwelling system), pipes, and lines, will potentially reflect, absorb, refract, or diffract a significant portion of the wave energy that impinges on the farm. The wave energy necessary to provide for upwelling may approach or exceed (during certain seasons) the total wave energy available to the farm (Dynatech 1978). A "wave shadow" may occur in the lee of the farm which could affect nearby coastal areas. Beaches and coastlines are complex entities created and dynamically maintained by wave energy and longshore currents (Shephard 1973: Bascom 1964; Donovan 1977). Any significant alteration of the energy content or angle of impingement of waves along a coastline would have substantial, although rather unpredictable, consequences and would affect both shoreline morphometry and littoral zone ecology. Beaches, bays, and offshore bars could unpredictably form or disappear in response to alterations in littoral energy patterns.

A large farm system could also alter surface current structure to an unknown, but potentially serious, extent. It has been predicted that even a small kelp farm (about 30 m diam) can theoretically induce a 90° change in local surface current direction (Oceanic Services 1977). A large kelp farm could alter surface current patterns over an extensive area. Currents influence beach structure and also act as delimiters of oceanic water masses, the characteristics of which are climatologically and ecologically important in coastal areas (King 1965; Sverdrup et al. 1942). A substantial change in oceanic current patterns could have far-reaching effects of great physical, meteorological, and biological magnitude.

These postulated oceanic kelp farms will probably rely on deep, nutrient-rich waters artificially upwelled to the surface for kelp fertilization. The volume of upwelled water required is impressive, and this large-scale movement of deep, cool water to the surface could both lower surface temperatures (which could affect near- and far-field weather, Wendland 1977) and disrupt the natural local and far-field thermocline structure. Upwelling sites along the Pacific coast are notorious areas of fog formation, where warm, wet oceanic air cools belows its dew point. Formation of large fog banks associated with upwelling for large kelp farms (Dynatech 1978) could lower kelp productivity (through light attenuation), interfere with navigation, and locally increase the earth's albedo.

5.5.2 Chemical Impacts

A number of potential chemical impacts exist associated with open-ocean kelp farming. Various toxic chemicals could leach from the support structure itself. Toxic compounds used in antifouling paints (usually copper) are designed to slowly leach into the surrounding water. Toxic organic chemicals (polychlorinated biphenyls, aromatic monomers, aldehydes such as formaldehyde, etc.) may be released from the degradation of synthetic support-grid and buoy lines and polyvinyl chloride piping (Hurby 1978). Although the release of toxic materials will be gradual, the vast amounts of synthetic materials necessary for the construction of these kelp farms [about 21,000 metric tons of plastic for the 40,470-ha farm (Wilcox et al. 1976)] could significantly affect local biota. Additionally, the ubiquitous problem of fouling in the open ocean has not been addressed. If fouling control is required to maintain the efficiency in the upwelling pipes, pumping mechanisms, and especially the distribution pipes (as will most likely be the case), it could add to the chemical insult of adjacent waters. The necessity for augmentation of the upwelling system by the use of municipal waste materials or fertilizer has not been defined, but, if used, could lead to chemical alterations of surface waters. Additionally, pest species that become established in the kelp might have to be controlled chemically, also adding to the alteration of the local chemical environment.

Seaweeds have been shown to exude toxic organic substances, such as phenols (Sieburth and Jensen 1969), and tannins (Sieburth and Conover 1965). For algae the rate of release of such material can be quite high. Phaeophyte tannins are considered antifouling agents for some brown algae (Sieburth and Conover 1965) and have been shown to suppress barnacle and muscle growth in tide pools dominated by certain brown algae (Conover and Sieburth 1966). Likewise, exudates from brown seaweeds are toxic to plaice larvae (Sieburth and Jensen 1969), plankton (Conover and Sieburth 1966), and bacteria (Scotten 1971 as cited in Hurby 1978). Production of such exudates by the kelp may affect water chemistry and locally alter the marine environment (Sieburth and Jensen 1969). Hurby (1978) has roughly estimated that phenol concentrations in water surrounding kelp farms could reach 1 ppm, which, if correct, might have detrimental effects on some open-ocean biota unadapted to their presence.

Two potential chemical impacts are associated with the upwelling of massive amounts of deep oceanic water: (1) alteration of the chemical composition of surface waters and (2) release of carbon dioxide into the atmosphere. Deep oceanic water masses are quite different from overlying surface waters. In the Pacific they are usually lower in oxygen but higher in nitrates, phosphates, carbon (dissolved CO_2, H_2CO_3, HCO_3^- and CO_3^{2-}), and some trace elements. Jackson (1977) reports that phytoplankton (diatoms) kept in deep ocean water were growth inhibited before addition of chelators to bind trace elements. The controlling factor appeared to be the free copper ion concentration. The upwelling of large amounts of deep ocean water may increase concentrations of trace metals at the surface, perhaps locally altering the planktonic species composition. On balance, however, it must be noted that trace metals are known to be adsorbed onto kelp tissue, which may aid in detoxifying the upwelled water (Cowen et al. 1976 as cited in Jackson 1977).

The cultivation of biomass for energy is attractive given a desire to decrease atmospheric CO_2 loading (Sect. 6.5). No more than 1 g of carbon is released to the atmosphere (as CO_2) during combustion of biomass (or biomass-derived products) for each gram of carbon originally fixed by the plant. On balance, no excess CO_2 would be liberated to the atmosphere to potentially enhance a greenhouse effect. This closed system is true for terrestrial biomass systems; however, the effect of large-scale ocean kelp farming on the carbon cycle is not as clear. The exchange of CO_2 between the atmosphere and the oceans occurs naturally. The cool, deep ocean waters of the Pacific contain more CO_2 (in various forms) than do surface waters. The movement of large masses of cool, deep water to the surface initiates warming and a surface buffering effect that might release CO_2 to the atmosphere. Williams (1975) has estimated that about four times as much carbon, could be released to the atmosphere by the cultivation of marine biomass and subsequent combustion of the methane generated from the biomass as would be released from burning an equivalent amount of fossil-fuel methane. If ocean kelp farming were conducted on a large scale, such additional CO_2 releases could potentially aggravate global increases in atmospheric CO_2 and associated climatological perturbations.

5.5.3 Biological Impacts

Many of the potential impacts on the physical and chemical environment from deployment of open-ocean kelp farms could have serious biological consequences.

Considerable opposition will probably arise to the introduction of *Macrocystis pyrifera* into waters that are not within its natural distribution. Attempts to introduce *Macrocystis* to European waters off the coasts of Scotland and Brittany have met with strong international objection from the International Council for the Exploration of the Sea, among others, on the grounds that "*Macrocystis* was considered to be a biological menace that would bring about unpredictable changes in existing marine ecosystems" and "might eventually infest European coasts from North Cape to Gibraltar" (Franklin 1974). At least the initial deployment of the ocean kelp farm is therefore assumed to be off the west coast of the United States.

If surface currents and/or wave patterns are significantly altered by the presence of a large offshore kelp farm, major modifications to the littoral environment could occur. The morphometry of beaches, offshore bars, and embayments might be altered. The physical and chemical characteristics of longshore water masses could change. These effects could unpredictably combine to redefine the littoral ecology of the adjacent coast.

Toxic substances exuded by the kelp (tannins and/or phenols), materials that have leached from synthetic support structure, toxic substances brought to the surface by upwelling, and chemicals used for pest or fouling control in the upwelling system could have deleterious effects on nearby pelagic and planktonic communities.

Kelp fronds are often lost from the parent plant because of wave damage, grazing, senescence, and mechanical harvesting (harvesting losses estimated at 10 to 20%, Flowers 1978). Kelp farms will be sources of large amounts of kelp detritus that could sink to the substrate or float inshore. If we assume a standing crop of 76 metric tons/ha (wet weight), this would total about 3,000,000 metric tons of kelp for the entire 40,470-ha farm. If storm damage caused only 10% of this standing crop to be dislodged, and only 10% of this kelp wrack were to drift ashore, this would produce a beach loading of about 190 kg (420 lb) wet weight of kelp per meter over a 160-km (100-mile) stretch of shoreline. Kelp storm wrack could significantly load beaches and littoral areas and could carpet large areas of ocean bottom, significantly altering benthic ecological communities.

Upwelling cool waters would alter local surface temperatures and salinities. The deployment of larger kelp farms might cause the formation of barriers to the migration of pelagic fish and to the distribution of planktonic species (Marr 1962, Marti 1969 as cited in Hurby 1978, Wiebe et al. 1976). Organisms that cannot escape the upwelling intake flow [about 1.3 m/s (4 fps) (Wilcox 1976)] will be entrained, and many will perish from mechanical, temperature, and pressure insult. The depths to which these upwelling pipes reach are not devoid of biota. Fish, prawns, zooplankton, medusae, squid, and many other animals inhabit this mesopelagic region. A 40,470-ha (100,000-acre) upwelling kelp farm unit might entrain about 50 metric tons of organisms per day if the average organism density were 0.05 g/m^3. Hurby (1978) concludes that, extrapolated to a large-scale kelp multifarm system, entrainment could result in major changes in the mesopelagic ecosystem.

Oceanic environments are relatively sparsely populated, nutrient- and substrate-limited ecosystems. A kelp farm would alter this environment. Nutrient-rich upwelled water would support expanded plankton communities in adjacent areas (assuming toxic substances do not inhibit plankton growth). The kelp and associated farm structures would also provide substrate for numerous invertebrates. Fish would obtain their subsistence from either the kelp and associated invertebrate population or from the enriched plankton community. Many fish use natural kelp beds in California for shelter, for food, or as areas of reproduction (North and Hubbs 1968), and fish are often closely associated with drifting kelp masses (Mitchell and Hunter 1970). Additionally, pelagic fish such as barracuda, albacore, bonito, and tuna are often caught adjacent to kelp beds or floating kelp. Depending on siting characteristics and the biological requirements of each species, many of these fish may become established as permanent or semipermanent residents of kelp farms and adjacent areas. The conditions may be propitious for the establishment of commercially exploitable concentrations.

Concomitant with this will be the displacement of indigenous pelagic species, either through direct usurpation of space, alteration of the chemical/physical environment, or competition with introduced species. The ecologic structure of the area would certainly be disrupted. A detailed, site-specific analysis would be necessary to determine the seriousness of this alteration. Placement of the farm systems will be critical in determining impacts to existing fisheries, with inshore placement probably more deleterious. The establishment of large populations of fish dependent on the kelp farm for habitat and nourishment could also cause problems during decommissioning.

In summary, some of the potential problems associated with large oceanic kelp farms may be serious. It is quite difficult, however, to assess the relative probabilities of these impacts as farm design and deployment are uncertain and necessary baseline data are lacking. Some of the impacts discussed here may not be supportable. Some consequences of kelp farming may not have been identified. Models must be developed to predict effects on local and distant wave and current patterns. The effects of artificial upwelling on the thermocline structure, release of carbon dioxide to the atmosphere, and alteration of surface-water characteristics need to be defined. Climatological impacts must be investigated, and the magnitude of organism entrainment through upwelling systems must be estimated and translated into impacts on the deepwater community. What *is* clear, however, is that if large-scale ocean biomass farming is seriously being contemplated as a source of biomass for energy, research must be initiated in a timely manner to provide the data necessary for an accurate assessment of, and development of mitigation measures against, potential environmental impacts.

References

Anderson, B. F., and M. R. Pelton, 1976. "Movements, Home Range, and Cover Uses: Factors Affecting the Susceptibility of Cottontails to Hunting," pp. 525–35 in *Proc. 30th Ann. Conference, SE Assoc. Fish Wildlife Agencies*, Jackson, Miss.

Aubertin, G. M., and J. H. Patric, 1974. "Water Quality after Clear-Cutting a Small Watershed in West Virginia," *J. Environ. Qual.* **3**: 243–49.

Bascom, W., 1964. *Waves and Beaches*, Doubleday, New York.

Benemann, J. R., 1978. *Biofuels: A Survey*, ER-746-SR, Electric Power Research Institute, Palo Alto, Calif.

Benemann, J. R., et al., 1977. *Species Control in Large-Scale Algal Biomass Production: Final Report*, SAN/7405-77/1, U.S. Energy Research and Development Admin., Washington, D.C.

———, 1978. "Fuels from Microalgae Biomass," pp. 891–946 in *Proc. 2d Annual Symposium on Fuels from Biomass, Troy, N.Y., June 20–22*, U.S. Department of Energy, Washington, D.C.

Blymyer, M. J., and H. S. Mosby, 1977. "Deer Utilization of Clearcuts in Southwestern Virginia," *South. J. Appl. For.* **1**: 10–13.

Bormann, F. H., et al., 1974. "The Export of Nutrients and Recovery of Stable Conditions Following Deforestation at Hubbard Brook," *Ecol. Monogr.* **44**(3): 255, 277.

Bradley, D., and D. Stevens, 1978. *Wood for Energy: An Interim Report on the Whole-Tree Harvesting Experiment in South Duxbury (Vt.)*, Vermont National Resources Council and Vermont Department of Forests and Parks, Montpelier, Vt.

Brown, C. W., A. R. Gahler, and R. B. Marson, 1973. "Nutrient Losses after Clearcut Logging and Slash Burning in the Oregon Coast Range," *Water Res. Res.* **9:** 1450–53.

Burger, G. V., 1978. "Agriculture and Wildlife," in *Wildlife and America*, H. P. Brokaw, ed., Council on Environmental Quality, Washington, D.C.

Campbell, R. B., et al., 1979. "Crop Residue Requirements for Water Erosion Control in Six Southern States," *J. Soil Water Conserv.* **34:** 83–85.

Carta, K. L., and J. A. Bondurant, 1976. *Control of Sediments, Nutrients, and Absorbed Biocides in Surface Irrigation Return Flow*, EPA-600/2/76-237, Environmental Protection Agency, Office of Research and Development, Washington, D.C.

Carter, L. J., 1977. "Soil Erosion: The Problem Persists Despite Billions Spent on It." *Science* **198:** 409–11.

Clark, W. C., D. D. Jones, and C. S. Holling, 1978. "Patches, Movements and Population Dynamics in Ecological Systems: A Terrestrial Perspective," in *Spatial Pattern in Plankton Communities*, J. S. Steele, ed., Plenum Press, New York.

Conover, J., and J. Sieburth, 1966, "Effect of Tannins Excreted from Phaeopyta on Planktonic Animal Survival in Tide Pools," pp. 99–100 in *Proc. 5th International Seaweed Symposium, Halifax*, E. G. Young and J. L. McLachlan, eds., Pergamon Press, Oxford, England.

Council on Environmental Quality, 1978. *The Ninth Annual Report of the Council on Environmental Quality*, Washington, D.C.

Cowen, J. P., V. F. Hodge, and T. R. Folsom, 1976. "In Vivo Accumulation of Radioactive Polonium by the Giant Kelp *Macrocystis pyrifera*," *Mar. Biol.* **37:** 239–48.

Cox, G. W., and M. D. Atkins, 1975. "Agricultural Ecology," *Bull. Ecol. Soc. Am.* **56**(3): 2–6.

Cramer, O. P., 1974. "Air Quality, Fire-Hazard, and Esthetics," pp. F1–F51 in *Environmental Effects of Forest Residues Management in the Pacific Northwest—A State-of-Knowledge Compendium*, O. P. Cramer, ed., USDA For. Serv. Gen. Tech. Rep. PNW-24, Pacific Northwest Forest and Range Exp. Sta., Portland, Ore.

CSO International, Inc., 1978. *Cost Analysis of Microalgae Biomass Systems: Final Report—Design Assumptions and Engineering Plans for 100-Square-Mile System*, CSO International, Inc., Concord, Calif.

DiNovo, S. T., et al., 1978. *Preliminary Environmental Assessment of Biomass Conversion to Synthetic Fuels*, EPA-600/7-78-204, Industrial Environmental Research Laboratory, Cincinnati, Ohio.

Doering, O. C., III, 1977. "Future Competitive Demand on Land for Biomass Production," in *Symposium on Clean Fuels from Biomass and Wastes, Orlando, Fla., Jan. 25–28, 1977*, Institute of Gas Technology, Chicago, Ill.

Donovan, N. S., 1977. *Wave Energy and Beach Response for Three Southern Rhode Island Beaches*, M.S. thesis, University of Rhode Island, Kingston.

Duvigneaud, P., and S. Denaeyer-de Smet, 1970. "Biological Cycling of Minerals in Temperate Deciduous Forests," pp. 199–225 in *Analysis of Temperate Forest Ecosystems*, D. E. Reichle, ed., Springer-Verlag, New York.

Dynatech, Inc., 1978. *Cost Analysis of Aquatic Biomass Systems*, prepared for U.S. Department of Energy, Div. Solar Technology, HCP/ET-4000-78/1, 2.

Elliott, L. F., J. M. McCalla, and A. Waiss, 1978. "Phytotoxicity Associated with Residue Management," in *Crop Residue Management Systems*, Am. Soc. Agron. Spec. Publ. 31, Madison, Wis.

Evans, E. E., and R. N. Conner, 1979. "Snag Management," pp. 214–25 in *Management of North-Central and Northeastern Forests for Nongame Birds*, Workshop Proc., U.S. Dep. Agric. For. Serv., North Cent. For. Exp. Stn., St. Paul, Minn., 268 pp.

Evans, K. E., and D. K. Dawson, 1976. "The Breeding Bird Survey in Missouri, 1967–1975," *The Bluebird* **43:** 12–22.

Flowers, A., 1978 (Director, Gas Supply Research, GRI). Letter to R. Ward, Chief of Fuels, from Biomass System Branch, DOE. *Reviewer's Comments on Cost Analysis of*

Aquatic Biomass Systems, HCP/ET-400-78/2, 110 pp.

Fogg, G. E., 1975. *Algal Culture and Phytoplankton Ecology*, University of Wisconsin Press, Madison, Wis.

Fortescue, J. A. C., and G. G. Marten, 1970. "Micronutrients: Forest Ecology and Systems Analysis," pp. 173-98 in *Analysis of Temperature Forest Ecosystems*, D. E. Reichle, ed., Springer-Verlag, New York.

Franklin, A., 1974. "Giant Kelp for Europe?" *New Sci.* **64**(927): 812-13.

Fredriksen, R. L., D. G. Moore, and L. A. Norris, 1975. "Impact of Timber Harvest, Fertilization and Herbicide Treatment of Streamwater Quality in Western Oregon and Washington," pp. 283-313 in *Forest Soils and Land Management, Proc. 4th North American Forest Soils Conference, Laval University*, B. Bernier and C. H. Winget, eds., Presses de l'Universite Laval, Quebec

Gagliano, S. M., and J. L. Van Beek, 1973. "Environmental Management in the Mississippi Delta System," *Gulf Coast Assoc. Geol. Soc. Trans.* **23**: 203-9.

Gikis, B. J., et al., 1978. *Preliminary Environmental Assessment of Energy Conversion Processes for Agricultural and Forest Product Residues*, EPA-600/7-78-047, NTIS PB-281 189, Municipal Environmental Research Laboratory, Cincinnati, Ohio.

Green, M. B., 1978. *Eating Oil—Energy Use in Food Production*, Westview Press, Boulder, Colo.

Gupta, S. C., et al., 1976. *Area Delineation of Possible Corn Residue Removal for Bioenergy in Four Minnesota Counties*, Minnesota Agric. Exp. Sta. Paper 9631, Minneapolis, Minn.

Hagen, L. J., and N. P. Woodruff, 1973. "Air Pollution from Duststorms in the Great Plains," *Atmos. Environ.* **7**: 323-32.

______, 1975. "Particulate Loads Caused by Wind Erosion," *Air Pollut. Control Assoc.* **25**: 860-61.

Hamrick, W. J., and J. R. Davis, 1972. "Summer Food Items of Juvenile Wild Turkeys," pp. 85-89 in *Proc. 25th Ann. Conf., SE Assoc. Game Fish Comm.*, Charleston, S.C.

Hart, C. W., Jr., and S. L. H. Fuller, 1974. *Pollution Ecology of Freshwater Invertebrates*, Academic Press, New York.

Haugen, A. O., 1943. "Management Studies of the Cottontail Rabbit in Southwestern Michigan," *J. Wildl. Manage.* **7**: 102-19.

Henry, J. F., M. D. Frazer, and C. W. Vail, 1976. "The Energy Plantation: Design, Operation, and Economic Potential," in *Thermal Uses and Properties of Carbohydrates and Lignins*, F. Shatizaden, K. V. Sankanen, and D. A. Tillman, eds., Academic Press, New York.

Holt, R. F., 1979. "Crop Residue, Soil Erosion, and Plant Nutrient Relationships," *J. Soil Water Conserv.* **34**: 96-98.

Howlett, K., and A. Gamache, 1977*a*. *Silvicultural Biomass Farms*, vol. 2. *Biomass Potential of Short-Rotation Farms*, Mitre Corp. Tech. Rep. MTR 7347, McLean, Va.

______, 1977*b*. *Silvicultural Biomass Farms*, vol. 6. *Forest and Mill Residues as Potential Sources of Biomass*, Mitre Corp. Tech. Rep. 7347, McLean, Va.

Hurby, T., 1978. *Primary Impacts of Growing Aquatic Plants for Energy*, WHOI 78-31, Woods Hole Marine Policy and Ocean Management Program, Woods Hole, Mass.

Hynes, H. B. N., 1971. *The Biology of Polluted Waters*, University of Toronto Press, Toronto, Ontario, Canada.

Jackson, G. A., 1977. "Biological Constraints on Seaweed Culture," in *Biological Solar Energy Conversion*, A. Mitsui et al., eds., Academic Press, New York.

Jemison, G. M., and M. S. Lowden, 1974. "Management and Research Implications," in *Environmental Effects of Forest Residues Management in the Pacific Northwest*, O. P. Cramer, ed., For. Serv. Gen. Tech. Rep. PNW-24, Pacific Northwest Forest and Range Exp. Sta., Portland, Ore.

Jones, J. L., 1979. "The Cost of Methane from Kelp Grown in an Open-Ocean Farm," *ASC/CSJ Chemical Congress Symp., Honolulu, April 1979.*

JPR Associates, Inc., 1975. *The Feasibility of Generating Electricity in the State of Vermont Using Wood as a Fuel: A Study*, submitted to Vermont Agency of Environmental Conservation, Department of Forests and Parks, Montpelier, Vt.

King, C. A. M., 1965. *An Introduction to Oceanography*, McGraw-Hill, New York.

Kockenderfer, J. N., and G. M. Aubertin, 1975. "Fernow Experimental Forest, West Virginia," pp. 14-24 in *Municipal Watershed Management, Symposium Proceedings,* U.S. For. Serv. Gen. Tech. Rep. NE-13, U.S. Department of Agriculture, Washington, D.C.

Kroodsma, R. L., and H. H. Shugart, 1975. *Effects of Whole-Tree Harvesting on Appalachian Mountain Bird Populations,* pp. 45-57 in Brooks, M. L., C. S. Hall, and J. Luchok (eds.), *Proc. Logging Residue Conf.*, West Virginia Office of Pubs., Morgantown, W. Va.

Larcher, W., 1975. *Physiological Plant Ecology,* Springer-Verlag, Berlin.

Larson, W. E., 1979. "Crop Residues: Energy Production or Erosion Control?" *J. Soil Water Conserv.* **34:** 74-76.

Larson, W. E., R. F. Holt, and C. W. Carlson, 1978. "Residues for Soil Conservation," in *Crop Residue Management Systems*, Am. Soc. Agron. Special Publ. 31, Madison, Wis.

Lecuyer, R., and J. E. Marten, 1976. "An Economic Assessment of Fuel Gas from Water Hyacinths," pp. 267-86 in *Symposium Papers—Clean Fuels from Biomass, Sewage, Urban Refuse, and Agricultural Wastes, Orlando, Fla., Jan. 27-30, 1976*, Institute of Gas Technology, Chicago, Ill.

Leopold, A. S., 1978. "Wildlife and Forest Practice," Chap. 8, pp. 108-20, in *Wildlife and America—Contributions to an Understanding of American Wildlife and Its Conservation*, H. P. Brokaw, ed., Council on Environmental Quality, Washington, D.C.

Levno, A., and J. Rothacher, 1967. *Increases in Maximum Stream Temperatures after Logging in Old-Growth Douglas-Fir Watersheds*, Res. Note PNW-65, Pacific Northwest Forest and Range Exp. Sta., Portland, Ore.

Likens, G. E., F. H. Bormann, and N. M. Johnson, 1969. "Nitrification: Importance to Nutrient Losses from a Cut-over Forested Ecosystem," *Science* **163:** 1205-6.

Likens, G. E., et al., 1978. "Recovery of a Deforested Ecosystem," *Science* **199:** 492-96.

Lindstrom, M. J., et al., 1979. "Tillage and Crop Residue Effects on Soil Erosion in the Corn Belt," *J. Soil Water Conserv.* **34:** 80-82.

Lockeretz, W., 1978. "The Lessons of the Dust Bowl," *Am. Sci.* **66:** 560-69.

Loehr, R. C., 1974. *Agricultural Waste Management—Problems, Processes, and Approaches,* Academic Press, New York.

Marr, J. S. W., 1962. "The Natural History and Geography of the Antarctic Krill (*Euphausia surperba* Dana)," *Discovery Rep.* **32:** 433-36.

Marti, Y. Y., 1969. "Marine Biological Resources of the Antarctic. First Expedition of R. V. Academician Knipowich," in *Proc. All Union Research Institute of Mar. Fish. and Oceanogr.*, vol. 66.

Megahan, W. F., 1972. "Logging, Erosion, and Sedimentation—Are They Dirty Words?" *J. For.* **40**(7): 403-7.

Mitchell, C. T., and J. R. Hunter, 1970. "Fishes Associated with Drifting Kelp, *Macrocystis pyrifera*, off the Coast of Southern California and Northern Baja California," *Calif. Fish Game* **56**(4): 288-97.

Morris, G. R., W. J. Jewell, and G. L. Casler, 1975. "Alternative Animal Waste Anaerobic Fermentation Designs and Their Costs," in *Energy Agriculture and Waste Management*, W. J. Jewell, ed., Ann Arbor Science Publications, Ann Arbor, Mich.

Morrison, I. K., and N. W. Foster, 1979. "Biomass and Element Removal by Complete-Tree Harvesting of Medium Rotation Forest Stands," pp. 111-29 in *Impact of Intensive Harvesting on Forest Nutrient Cycling*, State Univ. of New York, College of Environmental Science and Forestry, Syracuse.

North, W. J., and C. L. Hubbs, 1968. *Utilization of Kelp-Bed Resources in Southern California*, Calif. Dep. Fish Game Bull. 139, Sacramento, Calif.

Oceanic Services, Inc., 1977. *A Study of the Mechanism of Nutrient Dispersion in the Kelp Farm*, OSI 04589, Oceanic Services, Inc., Santa Barbara, Calif.

Onstad, C. A., and M. A. Otterby, 1979. "Crop Residue Effects on Runoff," *J. Soil Water Conserv.* **34:** 94-96.

Oregon State University, School of Forestry, 1977. *Silvicultural Chemicals and Protection of Water Quality*. EPA 910/9-77-036, NTIS PB 271 923, Environmental Protection Agency, Region 10, Seattle, Wash.

Oswald, W. J., and J. R. Benemann, 1977. "A Critical Analysis of Bioconversion with Microalgae," pp. 379-96 in *Biological Solar Energy Conversion*, Academic Press, New York.

Parr, J. F., and R. I. Papendick, 1978. "Summary—Crop Residue Management Systems: New Perspective for Soil, Water, and Energy Conservation," in *Crop Residue Management Systems,* Am. Soc. Agron. Special Publ. 31.

Pimentel, D., et al., 1973. "Food Production and the Energy Crisis," *Science* **182:** 443-49.

———, 1975. "Energy and Land Constraints in Food Protein Production," *Science* **190:** 754-61.

———, 1976. "Land Degradation: Effects on Food and Energy Resources," *Science* **194:** 149-55.

Raphael, M. G., and M. White, 1978. "Snags, Wildlife, and Forest Management in the Sierra Nevada," *California-Nevada Wildlife* **9:** 23-41.

Reid, G. K., 1961. *Ecology in Inland Waters and Estuaries*, Van Nostrand Reinhold, New York.

Reinhart, K. G., A. R. Eschner, and G. R. Trimble, Jr., 1963. *Effect on Streamflow of Four Forest Practices in the Mountains of West Virginia*, For. Serv. Res. Pap. NE-1, U.S. Department of Agriculture, Washington, D.C.

Salo, D. J., et al., 1977. *Silvicultural Biomass Farms*, vol. 3. *Land Suitability and Availability*, MTR-7347, The Mitre Corp., McLean, Va.

Segelquist, C., and M. Rogers, 1975. "Use of Wildlife Forage Clearings by White-Tailed Deer in the Arkansas Ozarks," *Proc. Ann. Conf. SE Assoc. Game Fish Comm.* **28:** 568-73.

Shephard, F. P., 1973. *Submarine Geology*, Harper and Row, New York.

Sieburth, J., and J. Conover, 1965. "*Sargassum* tannin, an Antibiotic Which Retards Fouling," *Nature* **208**(5005): 52-53.

Sieburth, J., and A. Jensen, 1969. "Studies on Algal Substances in the Sea. II. The Formation of Gellestaft (Humic Materials) by Exudates of Phaeophyta," *J. Exp. Mar. Biol. Ecol.* **3**(3): 275-89.

Skidmore, E. L., M. Kumar, and W. E. Larson, 1979. "Crop Residue Management for Wind Erosion Control in the Great Plains," *J. Soil Water Conserv.* **4:** 90-93.

Stone, E., 1973. "The Impact of Timber Harvest on Soils and Water," pp. 427-67 in *Report of the President's Advisory Panel on Timber and the Environment*, Government Printing Office, Washington, D.C.

Sverdrup, H. V., M. W. Johnson, and R. H. Flemming, 1942. *The Oceans—Their Physics, Chemistry and General Biology*, Prentice-Hall, Inc., New York.

Thornton, T., 1978. *An Assessment of Anaerobic Digestion in U.S. Agriculture*, ESCS-06, Economics, Statistics and Cooperatives Service, U.S. Department of Agriculture, Washington, D.C.

Tillman, D. A., 1978. *Wood as an Energy Resource*, Academic Press, New York.

(USDA) U.S. Department of Agriculture, Forest Service, 1971. *Wildlife Habitat Management Handbook*, Southern Region, U.S. Forest Service, Atlanta, Ga.

———, 1977. *Potential Cropland Study*, Soil Conservation Service, Statistical Bull. 578, Washington, D.C.

———, 1979. *Improving Soils with Organic Wastes*, 0-623-484/770, U.S. Government Printing Office, Washington, D.C.

U.S. Environmental Protection Agency, 1972. *The Relationship Between Animal Wastes and Water Quality*, President's Water Pollution Control Advisory Board, Washington, D.C.

———, 1978. *Agricultural Lands Protection Policy*, EPA Office of the Administration, Washington, D.C.

Van Doren, D. M., and R. R. Allmaras, 1978. "Effect of Residue Management Practices on the Soil Physical Environment, Microclimate, and Plant Growth," in *Crop Residue Management Systems*, Am. Soc. Agron. Special Publ. 31.

Varani, F. T., and J. J. Buford, Jr., 1977. "The Conversion of Feedlot Wastes into Pipeline Gas," in *Fuels from Waste*, L. L. Anderson and D. A. Tillman, eds., Academic Press, New York.

Vitousek, P. M., et al., 1979. "Nitrate Losses from Disturbed Ecosystems," *Science* **204:** 469-74.

Waide, J. B., and W. T. Swank, 1977. "Simulation of Potential Effect of Forest Utilization on the Nitrogen Cycle in Different Southeastern Ecosystems," pp. 767–89 in D. L. Correll, ed., *Proceedings of a Workshop on Watershed Research in Eastern North America*, Chesapeake Bay Center for Environmental Studies, Edgewater, Md.

Wayne, C. J., 1976. "The Effects of Sea and Marsh Grass on Wave Energy," *Coastal Res. Notes* **4**(7): 6–8.

Wells, C. G., and J. R. Jorgensen, 1979. "Effect of Intensive Harvesting on Nutrient Supply and Sustained Productivity," pp. 212–30 in *Impact of Intensive Harvesting on Forest Nutrient Cycling*, State Univ. of New York, College of Environmental Science and Forestry, Syracuse.

Wendland, W. M., 1977. "Tropical Storm Frequencies Related to Sea Surface Temperatures," *J. Appl. Meteor.* **16**: 477–81.

Wiebe, P., et al., 1976. "Gulf Stream Cold Core Ring: Large Scale Interaction Sites for Ocean Planktonic Communities," *Deep Sea Res.* **23**: 695–710.

Wilcox, H. A., 1976. "Ocean Farming," in *Conference on Capturing the Sun Through Bioconversion, Washington, D.C., Mar. 10-12, 1976,* Center for Metropolitan Studies, Washington, D.C.

Wilcox, H. A., et al., 1976. *Ocean Food and Energy Farm Project*, vols. 1-7, ERDA/USN/1027-76, prepared for U.S. Energy Research and Development Admin. by U.S. Naval Undersea Center, San Diego, Calif.

Williams, R. H., 1975. *The Greenhouse Effect and Ocean-Based Solar Energy Systems*, Center for Environmental Studies working paper 21, Princeton University, N.J.

6

Potential impacts of large-scale conversion

As described in Sect. 4.5, full-scale use of biomass for energy would involve a wide variety of processes and technologies. The processes would be deployed in various-size operations, depending on the local circumstances of feedstock availability, the form of energy desired, and the applicability of the technology to a particular scale of deployment. The current early stage of development makes it difficult to envision the eventual configuration of biomass energy systems and creates difficulties in accurately projecting the associated environmental impacts. In some cases the approximate emissions of conversion technologies can be specified, but the subsequent impacts of these emissions will be dependent on regional and local circumstances. In this discussion, therefore, impacts of biomass conversion can be considered only in a generic manner.

Several studies performed to date (e.g., Benemann 1978, Gikis et al. 1978, and DiNovo et al. 1978) have considered biomass conversion technologies and have made preliminary comparisons of environmental impact. Although these studies have been useful, a bias toward larger-scale systems has existed, perhaps because process descriptions and emissions data are more readily available for these units. In this book, medium- and small-scale systems have also been described (Sect. 4.5) because such units are likely to be commercialized and their impacts may be markedly different. In cases in which adequate data are unavailable to characterize the emissions and generic environmental impacts of biomass conversion systems, we have noted needs for additional assessment and research.

6.1 Combustion

A wide variety of biomass feedstocks are suitable for combustion, including wood chips, wood waste, cordwood, bagasse, and crop residues containing less than 50% moisture. Because of its familiarity and relative simplicity, direct combustion will undoubtedly be the use for a large amount of the suitable biomass.

The emissions from biomass combustion depend on many interacting factors (Sect. 4.4) such as particle size, moisture content, temperature of combustion, air-to-fuel ratio, size and shape of the combustion chamber, and mineral content of the feedstock. Because combustion conditions and control measures will depend to a great extent on the size of the units, the scale at which biomass combustion is deployed will be a critical factor in determining emissions and ultimate impacts.

6.1.1 Large-Scale Combustion

The effluent streams and land requirements of large-scale biomass combustion are characterized in Table 4.27. This table summarizes both the results of monitoring at the Moran Generating Station in Burlington, Vermont (described in Sect. 4.5.1.1), and the analysis of a hypothetical 3000-ton/d woodburning facility envisioned by Gikis et al. (1978).

The ash produced at a large woodburning power plant would probably be used in a landfill or perhaps as a soil supplement. Wood ashes contain nutrients required for plant growth, predominantly calcium, potassium, phosphorus, magnesium, sulfur (macronutrients) and sodium, silicon, iron, and other micronutrients. Use of ashes as a soil supplement would return some nutrients to soils. However, nitrogen is lost in combustion, and other nutrients are not readily available for plant uptake because of partial fusion of the ash (Calef 1976 citing King et al. 1965). The economics of returning the ash to forest lands for use as a soil supplement have not been analyzed, and a question exists as to whether or not using ash would be cost effective in maintaining soil fertility. Because of the unknown costs and unfamiliar procedures for large-scale spreading of wood ash, disposal in landfills seems likely, at least initially.

Although wood ash is a generally innocuous material, studies are needed to determine its leachability and any hazards that might result from its placement in landfills or on forest soils. Ash from woodburning power plants may contain materials not usually found in wood ash. For instance, wood is often cofired with oil, and at the Moran station, wood chips were also contaminated with coal fines from the adjacent storage pile (Sect. 4.5.1.1). In that study polycyclic aromatic hydrocarbons (PAHs) were found in both the bottom ash and the collected fly ash. It is not known what concentration of PAH would result in the fly ash from burning of wood chips alone or if PAH production is increased or decreased by the combustion of an oil-wood combination. If wood ash is disposed of in a landfill, phosphorus, trace elements, and PAHs could be leached from disposal sites and enter groundwater. The Environmental Protection Agency (EPA) assessment of wood ash from the Moran station (Golembiewski 1978) compared the concentration of elements in bottom ash and fly ash with concentrations considered to be hazardous. This scheme identified the following elements as causing the greatest environmental concern: bottom ash—P, Mn, Fe, Ni, Li, Cr, V, Be, Cu, Co; fly ash—P, Mn, Fe, Se, Ni, Li, As, Cr, Cu, Al. However, the EPA assessment methodology provides only a crude indication of potential hazard because the chemical form of elements and the leachability of the ash were not considered (Golembiewski 1978). The mobility of phosphorus, trace elements, and PAHs in landfills or forest soils depends on many factors, and further research is needed in this area.

Aside from cooling water required for any power plant, the only aqueous effluents likely to be associated with wood combustion are leachate and runoff from wood-chip storage piles and condensate from chip-drying operations. Runoff effluents have not been characterized but are expected to have moderate to high biological oxygen demand (BOD) loading (100 to 1000 mg/liter) resulting from organic compounds leached from wood and an acidic leachate (Sect. 4.1.2.1 and Table 4.4). Condensate from wood drying would probably be routed to the plant's sanitary waste treatment system or to the municipal sewer. Assuming that the organics contained in the wastewater were readily decomposed by biological treatment, no adverse impacts should result from such a waste stream (see also Sect. 4.4.2.2).

The principal air emissions from large-scale combustion are particulates, CO, hydrocarbons, and NO_x (Table 4.22). Although burning wood instead of coal might reduce the adverse environmental impacts of SO_2 and possibly NO_x emissions, emissions of particulates, hydrocarbons, and carbon monoxide from wood burning might equal or exceed emissions from coal burning (Gikis et al. 1978). Of course, if wood replaced a relatively clean fuel such as natural gas or low-sulfur oil, increased emissions and adverse impacts to air quality would be expected. The ultimate impacts of air emissions from biomass-fueled power plants would vary greatly, depending on the efficiency of controls, local topography, meteorological conditions, and ambient air quality.

The tendency for high particulate emissions from wood combustion is well known. Gikis et al. (1978) believe that large plants with emissions control equipment designed to capture particles produced by burning wood could meet EPA's new source performance standards for particulates of 50 $g/10^9$ J (0.1 $lb/10^6$ Btu) output. However, smaller wood-fired boilers might have difficulty meeting these standards (Benemann 1978). Particulate emissions from wood combustion could cause reduced visibility and aesthetic degradation; effects on human health are poorly known. Although particulates from large-scale wood combustion could adversely affect air quality, the greatest impacts would result from small, poorly controlled units (Sect. 6.1.2).

As with particulate emissions, carbon monoxide (CO) and hydrocarbon (HC) emissions from large-scale wood combustion can be high if not controlled (Sect. 4.4.1). For instance, it may be difficult to control emissions from boilers that are converted to burning biomass from fossil fuels, especially oil- or gas-fired boilers. However, any new woodburning power plants or large biomass-fired boilers designed for highly efficient and complete combustion should have reduced CO and HC emissions. In general, CO emissions from stationary sources have been considered of minor importance because space heating and power generation produce less than 2% of the total CO emissions, whereas transportation produces 77% of the total, principally from internal combustion engines (National Academy of Sciences, 1977*a*). However, large-scale biomass-fired boilers have the potential for degrading local air quality through increases in CO and HC emissions. Their emission rates and ultimate impacts depend largely on the plant design, control equipment, and operating procedures. Small-scale combustion units (fireplaces and woodstoves) seem to have far greater potential for greatly increasing emissions of CO and HC, and the potential impacts of these pollutants are discussed in Sect. 6.1.2.

As noted in Sect. 4.4.1.1, NO_x emissions from biomass combustion can be thermally produced from N_2 in air or can derive from the nitrogen contained in biomass. Under certain circumstances both types of NO_x emissions could cause problems. Because most of the biomass materials that will be burned contain less than 0.5% nitrogen (Sect. 4.1.2) compared with an average of 1.3% for coal (Ruch et al. 1974), NO_x emissions would not appear to be as serious a problem as emissions from fossil fuel combustion. However, a biomass fuel such as bark containing 59% moisture and 0.3% nitrogen, as received, with an effective heating value of 11.5×10^6 J/kg (5000 Btu/lb) theoretically could yield 300 g/10^9 J (0.6 lb/10^6 Btu) if only 50% of the fuel nitrogen is converted into NO (Sect. 4.4.1.1). Anything above this emission rate would exceed the proposed EPA standard for NO_x of 300 g/10^9 J (0.6 lb/10^6 Btu) input for combustion in new electric steam generating units that use solid fuels other than coal (U.S. Environmental Protection Agency 1978*a*).

Thermally produced NO_x emissions may be a problem in cases where biomass combustion causes very high combustion temperatures. As noted in Sect. 4.4.1.1, cofiring equal quantities of coal and wheat straw caused NO_x emissions to increase fourfold over firing coal alone. Since materials such as sawdust, pelletized wood, dry wood chips, and dry agricultural residues may produce intensely hot combustion, NO_x emissions from burning such materials should be quantified. If NO_x emissions increase as a result of biomass combustion, there may be impacts to human health, vegetation damage, and ecosystem-level effects. Exposure to NO_x causes reduced resistance to respiratory infection, and NO_x is thought to act synergistically with other air pollutants to cause chronic lung disease. Also, oxides of nitrogen may cause ecological impacts through their participation in the production of ozone and peroxyacylnitrates, which are damaging to vegetation (National Academy of Sciences 1977*b*).

6.1.2 Medium- and Small-Scale Combustion

Wood ash from small- and medium-scale combustion units is very likely to be used as soil supplements in gardens or lawns. Ashes may also be disposed of in backyards and in municipal solid waste systems. No adverse impacts are anticipated from this practice.

A noteworthy aspect of home woodburning is the potential hazard from creosote formation. The black tarlike substances that condense in flue pipes are the product of pyrolysis and combustion. Stoves with high energy efficiencies are more likely to have chimney creosote problems because they do such a good job of extracting heat from the flue gases (Shelton 1976). In addition to creating hazards from chimney fires (if flue pipes are not cleaned regularly), creosote allows potentially hazardous exposure to compounds produced in pyrolysis. As discussed in Sect. 6.2, these substances may pose a hazard because of possible toxicity and carcinogenicity.

Assessing the air quality impacts of small-scale combustion systems is difficult, if not impossible, by use of classical techniques for determining air quality. Current air quality studies typically deal with the impacts associated with large, centralized emission sources. These sources have emission standards that are strictly enforced and commonly employ sophisticated emission abatement devices to ensure compliance with standards. The fuels used by these sources are uniform, and their combustion devices are carefully designed, operated, and maintained. When area sources of pollution are investigated by use of classical techniques, uniform fuels and combustion efficiencies can be assumed with some degree of confidence. Unfortunately, none of these assumptions are valid when small-scale biomass conversion systems are considered. The fuels used vary in heat content and pollution potential. The combustion devices have not been designed, operated, or maintained for maximum efficiency or environmental control, nor are they typically regulated when used in residences or small commercial operations.

Nonetheless, it is essential to attempt an estimate of the air quality impacts of small-scale combustion. If wood stoves replace more conventional energy sources, the resulting increase or decrease in air pollutant emissions becomes a most important consideration in the evaluation. Emissions factors compiled by the U.S. Environmental Protection Agency (1977) are available for conventional fuels. Air pollution emissions from the combustion of conventional fuels replaceable by biomass are compared with emissions from woodburning in Table 6.1. Clearly, replacing a clean fuel such as natural gas will result in greater net emissions, whereas replacing a more polluting fuel such as coal or oil will result in decreased emissions of sulfur, NO_x, and sometimes hydrocarbons.

Small-scale combustion of woody biomass results in the emission of air pollutants (primarily particulate matter), carbon monoxide, and hydrocarbon gases. To quantify potential impacts of direct woody biomass combustion, estimates of the ground-level concentrations of particulates and carbon monoxide from home woodburning stoves are calculated. The assumptions used in the analysis are listed in Table 6.2. The estimated concentrations 1000 m downwind of the area

Table 6.1. Emissions from residential heating fuels (pounds of pollutant per Btu $\times 10^6$)

Pollutant	Coal[a]	Fuel oil[b]	Natural gas[c]	Wood[d]
Particulates	0.8	0.02	0.005—0.015	0.5[e] 2.5[f]
Oxides of sulfur[g]	3.4	0.08	0.0004	0[h]
Oxides of nitrogen[i]	0.1	0.1	0.08—0.12	0.01[h]
Hydrocarbons[j]	0.8	0.01	0.02	0.05[h]
Carbon monoxide	0.5	0.04	0.02	16[h]

[a] Assumes bituminous coal, 2.3% sulfur, 13,000 Btu/lb.

[b] Assumes No. 2 fuel oil, 0.3% sulfur, 19,376 Btu/lb.

[c] Assumes 994 Btu/ft^3.

[d] Home woodburning stove; assumes 7180 Btu/lb for oak, 7960 Btu/lb for pine, and 50% efficiency.

[e] Oak.

[f] Pine.

[g] Expressed as SO_2.

[h] Same as for residential fireplace.

[i] Expressed as NO_2.

[j] Expressed as CH_4.

Source: U.S. Environmental Protection Agency, 1977. *Compilation of Air Pollutant Emission Factors*, 3d ed., EPA Rep. AP-42, Research Triangle Park, N.C.

Table 6.2. Assumptions used in the analysis of direct woody biomass combustion

Meteorological conditions	
Wind speed	2 m/s
Effective emission height	10 m
Atmospheric stability	Stable, class E
Study area	
Size	0.5 × 0.5 km
Housing density	4 units/acre, ≃ 247 dwellings
Emission	
Wood use	4 lb/h/dwelling
Particulates	0.05 lb/h/dwelling
CO	0.30 lb/h/dwelling

resulting from biomass combustion are 53 $\mu g/m^3$ for particulates and 340 $\mu g/m^3$ for CO, using the techniques suggested by Turner (1970). These predicted concentrations are for a short time period and would be less for an averaging time greater than a few hours. However, such concentrations could be significant when added to ambient levels. When considered in the context of current air quality regulations, such an increase in emissions from a regulated source might not be permitted in most areas of the country and would not be permitted near pristine air quality areas such as national parks or national wilderness areas. Increased emissions from small-scale combustion also raise the question of adverse effects on human health resulting from chronic exposure to CO and particulates.

Carbon monoxide has natural background levels between 0.03 and 0.1 ppm (Dreisbach 1974). The EPA standard for CO is 9 ppm maximum for 8 h average exposure and 35 ppm maximum for 1 h; the maximum allowable concentration for an exposure of 8 h/d, 6 d/week is 50 ppm (National Academy of Sciences 1977*a*). Cities with heavy vehicular traffic have been monitored at greater than 20 ppm for continuous periods as long as 24 d and in excess of 10 ppm for almost a year (Dreisbach 1974). The above analysis suggests that if 247 homes within a 0.25-km^2 (62-acre) area were heated with wood stoves, downwind air quality could temporarily suffer additional degradation. Elevated carbon monoxide levels are suspected of causing a variety of adverse health effects, including decreased ability to concentrate, aggravation of heart disease, and retardation of fetal growth (National Academy of Sciences 1977*a*). Further analysis of CO emissions from wood stoves and their effects are clearly needed.

The effects of particulates from biomass combustion on human health are poorly known. Cramer (1974) points out that wood smoke from wildfires has been present throughout human evolution, and there is no conclusive evidence that wood smoke causes permanent injury to human health (but see also Sect. 4.4.2.1). However, particulates from woodburning are thought to be hazardous because small particles act as sites for condensation of organic substances and volatile trace elements such as arsenic, mercury, and cadmium. Because particles from wood combustion are respirable and subject to deposition in the lungs, increased particulate emissions could lead to increased exposure to volatile elements and polycyclic organic matter (POM). Table 6.3 identifies some of the specific polycyclic organic compounds found in flue gases from biomass combustion. Although the amount of POM from residential combustion of wood is relatively small, typically less than 1% of the total condensable organic fraction, some of the individual POM compounds listed in Table 6.3 are known to be carcinogenic (Milliken 1979).

Research by the EPA has tentatively established a POM emission factor of 5×10^{-4} to 0.2 g/kg for residential wood combustion (Milliken 1979). The wide range of this emission factor reflects the great variation in combustion efficiency. The largest source of POM emissions results from inefficient combustion in hand-stoked residential coal furnaces (National Academy of Sciences 1972). POM emissions from woodstoves are expected to be smaller than those from residential coal furnaces (Milliken 1979). Nonetheless, investigation is needed into emission rates, composition, and health effects of particulates from wood combustion.

Airborne emissions from woodstoves have far greater significance when one considers the following factors: (1) woodstoves are entirely unregulated; (2) in certain regions, woodstoves are being adopted rapidly (Roop 1978), especially the airtight models, which can promote incomplete combustion under "slow-burning" conditions (Sect. 4.4.2.1); and

Table 6.3. Polycyclic organic compounds identified in biomass combustion flue gas[a]

Compound	Fireplace[b]	Wood combustion source: Laboratory wood burn[c,d]	Laboratory burn of pine needles[e]	Grass clippings, leaves, and branches[f]
Anthracene/phenanthrene	+	+	+	+
Methyl anthracenes	+	+	+	
Fluoranthene	+	+	+	+
Pyrene	+	+	+	+
Methyl pyrene/fluoranthene	+	+	+	
Benzo(*c*)phenanthrene	+		+	
Chrysene/benz(*a*)anthracene	+	+	+	
Methyl chrysenes	+	+	+	
Benzo fluoranthenes	+	+	+	
Benzo(*a*)pyrene	+	+	+	+
Benzo(*e*)pyrene	+	+	+	+
Perylene	—	+	+	+
Methylcholanthrenes	+			
Indeno(1,2,3-cd)pyrene	+	+	+	
Benzo(ghi)perylene	+	+	+	+
Dibenzo(a,h)anthracene	—	+		
Dibenzo(c,g)carbazone	—			
Dibenz(ai and ah)pyrenes	—	+		+
Coronene	—			—

[a]The plus sign indicates that the specific polycyclic organic compound was identified in the wood combustion emission; the minus sign indicates that the compound was analyzed for but not detected; a blank indicates either that the compound was not analyzed for or not present at detectable levels.

Source: Milliken, J. O., 1979. "Airborne Emissions from Wood Combustion," presented to the Wood Energy Institute Wood Heating Seminar IV, Mar. 22—24, Portland, Ore. Data compiled from:

[b]U.S. Environmental Protection Agency, 1975. *Source Sampling Residential Fireplaces for Emission Factor Development*, EPA-450/3-76-010.

[c]Lee, M. L., et al., 1977. "Source Identification of Urban Airborne Polycyclic Aromatic Hydrocarbons by GC/MS and HRMS," *Biomed. Mass Spec.* 4: 182—86.

[d]There are another 21 polycyclic organic compounds that were reported but not indicated on this table.

[e]McMahon, C. K., and S. N. Tsoukalas, 1978. "Polynuclear Aromatic Hydrocarbons in Forest Fire Smoke," in *Carcinogenesis*, vol. 3, *Polynuclear Aromatic Hydrocarbons*, P. W. Jones and R. I. Freudenthal (eds.), Raven Press, New York.

[f]Hangebrauck, R. P., D. J. vonLehmden, and J. E. Meeker, 1967. *Sources of Polynuclear Hydrocarbons in the Atmosphere*, U.S. HEW, Public Health Service, AP-33, PB-174-706, Washington, D.C.

(3) homeowners often wish to operate their stoves under conditions most conducive to formation of particulates and carbon monoxide (e.g., slow burning to keep a fire going overnight, burning green wood, etc.). In order for wood stoves to produce a significant amount of energy on a national scale, a great many stoves are required. About 30 million stoves would be required to produce 1 EJ (1.06 quad) of energy per year at 50% efficiency (assuming 2 cords of dry wood per household per year, 2 short tons per cord, and 8000 Btu per pound of dry wood). Thus the possibility of increased air emissions from home woodburning clearly requires further detailed research and assessment.

6.2 Thermochemical Conversion

Technologies for thermochemical conversion, including air gasification, oxygen gasification, and pyrolysis, are suitable for conversion of woody biomass, manure, and agricultural materials containing less than

50% moisture. The emissions and land requirements are generally summarized in Table 4.33. However, only preliminary characterization of the emissions of thermochemical conversion has occurred.

6.2.1 Large- and Medium-Scale Conversion

In some respects, the impacts of large- and medium-scale conversion facilities would resemble those of a small-scale coal gasification facility. For instance, solid wastes from thermochemical conversion would include ash, char, and perhaps sulfur from gas cleanup. Composition of the ash would reflect the mineral content of the feedstock, but organic substances would be present in char and/or ash as a result of incomplete processing. As with residuals from coal conversion (Braunstein et al. 1978), the chemical form of substances in ash or char from biomass is unknown (Sect. 4.1.2 and Table 4.4). Ash and char will probably be landfilled. Tests will be required to determine if this solid waste is considered "hazardous" under the criteria defined by the Resource Conservation and Recovery Act. The ash and char can likely be disposed of without adverse impacts, but further research is needed to characterize the waste and to determine the leachability of organics or trace elements.

Thermochemical processes generally include quench stages, separators, wet scrubbers, or condensers, which produce aqueous streams containing some of the organic compounds produced during conversion. Composition of this aqueous stream would vary with the feedstock (Sect. 4.5.2.1). Such aqueous waste streams are likely to have a high BOD, perhaps ranging from 500 to 3000 mg/liter (DiNovo et al. 1978). The substances contained in aqueous streams may be toxic or carcinogenic. Resin acids condensed from veneer-drying stack gas were toxic to young salmon at the 1- to 2-ppm level (Sect. 4.4.2.2). However, the condensed pitch from the inside of a dryer stack was only mildly toxic to mice with an LD_{50} of 1100 mg/kg injected intraperitoneally (Fraser and Swan 1972). Aqueous streams may require advanced physical-chemical treatment to avoid adverse impacts, since biological treatment systems may be disrupted by toxic constituents in the wastewater.

Air emissions would result from (1) combustion of produced gas and (2) other stack gases from the conversion processes. Emissions from combustion of other products (char, charcoal, or pyrolysis oil) that occur offsite are beyond the scope of this assessment. Because product gases can be scrubbed to remove particulates and acid gases (H_2S, CO_2), air emissions could be generally low. Also, because the gas produced has a relatively low heating value and flame temperature, NO_x emissions are expected to be low (DiNovo et al. 1978).

It is assumed that large-scale conversion facilities will be designed to employ state-of-the-art environmental control equipment and that all emissions will be controlled to acceptable levels.

6.2.2 Small-Scale Conversion

Small-scale thermochemical converters such as the Tennessee Valley Authority (TVA)–Maryville College pyrolysis unit (Sect. 4.5.2.3) are difficult to assess because of a lack of operating experience and emissions data. The unit is not expected to produce solid wastes or wastewater. Ash will be contained in the charcoal product, which may be burned or used for activated carbon. Water from the pyrolysis unit will travel as vapor with the product gases. Preliminary analysis indicates that air emissions from small units are not likely to cause serious adverse impacts (Hodam and Williams 1978). However, further research is needed to characterize the unplanned emissions (smoke and tar condensate) to identify applicable control measures and to determine the impacts of possible emissions.

A major issue surrounding thermochemical processes is the potential hazard posed by products of pyrolysis. Gikis et al. (1978) identified 22 potentially hazardous chemicals or chemical groups produced in carbonization of wood (based on research of Jahnsen 1961). These hazardous materials include dimethylamine, butyrolactone, catechol, pyrogallol, *o*-, *m*-, and *p*-cresol, estragole, limonene, pinene, chrysene, oleic acid, octadecane, eicosane,

3,4-benzpyrene, 1,2,5,6-dibenzanthracene, 20-methylcholanthrene, creosol, and xylenols. Some of these compounds are known, or suspected, to be carcinogens or tumor-promoting agents. Their presence and activity in pyrolysis oil are not yet known, and this is clearly an area for further research. Pyrolysis oil should be handled as a possible carcinogen, and the various streams and emissions from thermochemical biomass processing should be analyzed to determine the extent of the hazard.

6.3 Anaerobic Digestion

Microbial conversion of wet biomass materials (manures, aquatic plant matter, and high-moisture agricultural residues) produces a gaseous mixture, a liquid effluent, and a solid sludge. If kelp or other marine algae are used as feedstocks, they must be pretreated to remove salt, which results in an additional waste brine (U.S. Environmental Protection Agency 1978*b*). Use of manure, in particular, for methane production is one of the most promising near-term biomass technologies because of ample supplies of feedstock (Sects. 2.3, 3.3, and 5.3). Implementation of both commercial-scale and on-farm digester technology is currently feasible with animal wastes. Use of kelp or other novel feedstocks entails more uncertainty for commercial application.

6.3.1 Commercial Scale

The gas resulting from digestion is produced at a rate of about 7 to 12 m^3 per 100 kg (509 ft^3/lb) of manure or algae feedstock converted. Roughly 60% of the produced gas is methane and 30 to 35% is carbon dioxide. Although handling and storage of methane, an explosive gas, is well established in the natural gas industry, thorough training of the staff that operates compression and transmission equipment will be crucial for safety in operations where gas is upgraded (98% methane) for commercial sale. Digester gas almost always contains traces of hydrogen sulfide, which is toxic, corrosive, and has an offensive odor. Techniques for removing hydrogen sulfide are relatively simple but will entail additional costs (Thornton 1978). Other air emissions (particulates, oxides of nitrogen, and hydrocarbons) will be insignificant or minimal and should not require control equipment. Sulfur, however, appears to be a constituent of kelp that occurs in sufficient quantities to require attention (Sect. 4.1.8). (Sulfur is roughly three times as concentrated in kelp as it is in manure.) Preliminary calculations indicate that the output of a 100,000-acre kelp farm could include as much as 40,000 tons of sulfur annually (Sect. 4.4.5). Because little is known about the form in which this sulfur will appear, additional study is necessary to determine whether or not an impact mitigation strategy is required.

Poisoning of bacteria in the digester caused by the buildup of sulfide, ammonia, metals, or acids could occur occasionally. This poisoning would mean that "sour" digester contents that are only partially treated would have to be emptied out and disposed of. These materials are vile smelling because of low-molecular-weight volatile acids (butyric and propionic) and could cause odor problems downwind from a large plant and/or disposal site (landfill) (U.S. Environmental Protection Agency 1978*b*).

Liquid effluent from the digestion process would be largely recycled. Mixed digester residue (solids plus fluids) could be spread as a soil amendment on agricultural or other land (such as highway median strips and parklands). Digester residue could also be recycled to aquatic biomass production systems (Sect. 3.4). Up to 80% of the fluid from digesters can be reused by mixing it with biomass as it enters the digester. Such methods for use of digester residues requires attention to the fate of heavy metals, organic substances, and perhaps pathogens. Land spreading is presently used to dispose of digested sewage sludge. Adverse impacts to soils, groundwater, and drinking water supplies are avoided by limiting the rates of waste application, choosing good disposal sites, and monitoring.

If digester residues are recycled to production systems or digesters, heavy metals may build up in the system. Marine and aquatic plants bioaccumulate certain elements (Sects. 3.4 and 4.1.8), thus increasing this tendency. Recycling digester residue requires a bleed stream to prevent buildup of substances toxic to anaerobic organisms, such as ammonia, sodium, calcium, magnesium, and heavy metals. Water that is not recycled must be treated before discharge to sewers or application to land. The BOD of the liquid stream is high (500 mg/liter), and suspended solids may be greater than 700 mg/liter (U.S. Environmental Protection Agency 1978*b*). Liquid waste from brine-removal pretreatment of marine plants can be recycled to kelp beds or to other production sites to provide nutrients, if this procedure is economic. Otherwise, this liquid waste will also require secondary treatment. A large-scale facility that processes 320 metric tons of manure daily would consume about 1.2×10^5 to 2.5×10^5 m^3 (100 to 200 acre-ft) of water per year (Varani et al. 1977).

The dewatered solid sludge resulting from the digestion process is odor free, biologically stable, and unattractive to pests (Thornton 1978). The sludge has potential as fertilizer and can be handled in the same manner as regular manure. Sludge from both manure and kelp processing could be used as animal feed, especially in light of its high amino acid content [20 to 25% dry weight basis for manure; 37% crude protein in insoluble digester effluent from kelp (Thornton 1978, Electric Power Research Institute 1978)]. Continual refeed of manure-derived material to animals could eventually, however, lead to possible deleterious concentrations of trace elements such as arsenic (from chicken feed supplements) and pesticides in meat of fed animals, depending on the number of recycles. Because the trace element content of manure is five to ten times higher than that in animal feed (Sect. 4.1.7), it will be important to examine the possibility of magnification of toxic elements by continuous sludge refeeding (Capar et al. 1978).

Sludge may be stored onsite at large-scale facilities for further drying before final disposal by spreading. Six months of sludge storage (54,000 metric tons) for a 320-metric-ton/d manure-processing facility would require about 2 ha (5 acres) of storage area, if material were piled about 3 m high (Varani et al. 1977). Careful management and site selection for stored sludge will be needed to minimize leaching and runoff contamination problems similar to those with raw feedlot manures.

Data to completely characterize the residuals from anaerobic digesters do not yet exist. Information that should be gathered in the future includes data on organic matter, BOD, chemical oxygen demand (COD), bacteria and pathogen counts, pH, metals, and trace elements in both aqueous waste and solid residual. Leach tests on stored solids under expected environmental conditions should provide useful data on contaminant mobility potential.

6.3.2 On-Farm Facilities

Biogas produced from manure in smaller-scale, on-farm facilities will not need to be upgraded, as with gas for commercial pipelines. For use near the digester (e.g., on-farm heating of water or buildings), raw biogas is suitable and even preferable to minimize cost and effort. Hydrogen sulfide must still be removed, however, for protection from corrosion. Because other air emissions are expected to be minimal, on-farm facilities are unlikely to become a significant source of uncontrolled atmospheric pollutants. Liquid and solid waste products, although in smaller amounts, will still require handling as in large-scale facilities (i.e., lagooning, land spreading, dewatering for feed production, etc.).

6.4 Fermentation

The most direct production process for ethanol is by fermentation of sugar juices. At present, the most abundant and economical source of this material is sugarcane. Juice is mechanically extracted from cane, leaving a residual material (bagasse) that can be further dried and burned for mill boiler fuel. The effects of bagasse combustion on air quality are not unlike those of wood combustion (Sect. 6.1). If not burned, the bagasse must be disposed of as a solid waste. More commonly, however, bagasse provides an economical fuel from which electricity and/or process steam can be generated.

Fermentation of the pressed raw sugar mash produces ethanol, fusel oil, and "slops" or stillage. Stillage represents a fairly large quantity of solid waste. Data from the Brazilian gasohol program indicate that stillage is produced at a rate of 12 to 13 m^3 per cubic meter of alcohol. Thus, for example, if the entire annual production of sugar (6 million metric tons) were fermented, it would be necessary to dispose of 46 to 50 million cubic meters of stillage. This is not, however, as much waste as industrial processes typically produce, and it would be spread out over numerous locations. Stillage does have a high pollution potential because of its high soluble organics and inorganics content (Yang and Trinidade 1978), but such wastes are unlikely to be considered hazardous. If stillage is not treated, however, its strong odor will require that it be incorporated into the soil. Fermentation sludge may be valuable as a fertilizer if supplemented with the minerals it lacks.

Production of ethanol from corn and other grains requires a mild preliminary treatment to hydrolyze starch into sugars (Sect. 4.5.4). After conversion of starch to sugar, fermentation proceeds as with sugarcane mash. No bagasse-like material is available for firing mill boilers because the grain is harvested without stover; therefore, an external fuel source is needed. This may reduce net energy gain substantially. It is, nevertheless, a feasible method of converting more available fuels to a desirable liquid transportation fuel. Stillage from fermentation of corn and wheat has value as an animal feed supplement. Another possible by-product is brewer's yeast, a source of B vitamins used widely in the food industry for enrichment of white flours.

Conversion of wood or other highly cellulosic materials to fermentable sugars entails a more severe hydrolysis pretreatment and production of a sizable calcium sulfate sludge (Sect. 4.5.4 and Table 4.39). Use of wood chips, for example, requires treatment with live steam and sulfuric acid at high temperatures and pressures. Products include lignin (60 to 70% of the insoluble residue), methanol, acetic acid, furfural, nonfermentable sugars, and the desired fermentable sugars. Methanol and acetic acid may be recoverable by-products in commercial-scale operations. The liquid waste stream has a high BOD and requires secondary aerobic treatment or land treatment. Effluent produced in one study amounted to 7 m^3/metric ton (1600 gal/ton) of material hydrolyzed [230-metric-ton/d (250-ton/d) capacity (U.S. Environmental Protection Agency 1978*b*)]. Based on organic loads, a large-scale plant is projected to need a secondary treatment facility as large as that required by a medium-size town of 50,000 to 80,000 people. Dewatering of residual lignin, which has a high heating value, may permit its use as solid fuel. Carbonization or solvent extraction of lignin may also provide new sources of compounds currently synthesized from petroleum. The EPA considers acid hydrolysis of woody materials to be a technology that is not yet sufficiently attractive to warrant in-depth research; hence, its environmental impacts remain largely unquantified.

Hydrolysis of cellulosic substrates can also be accomplished by enzymatic hydrolysis. The advantages of the enzymatic process over acid hydrolysis are (1) enzymes react specifically with cellulose; (2) the process occurs under moderate conditions, reducing corrosion potential; and (3) the glucose product is relatively pure. The uncertainties of this method are, however, even greater than for acid hydrolysis, which is better established as a conversion process (U.S. Environmental Protection Agency 1978*b*).

In summary, constraints on providing fuels from fermentation appear to be more severe at the feedstock production phase (Sect. 5.2) than at the conversion phase of the fuel cycle. Although uncertainties exist, especially in the area of stillage disposal, the quantities of waste produced will probably be on a scale amenable to environmental control.

6.5 Biomass for Energy and the Global Carbon Cycle

Monitoring of the atmospheric concentration of CO_2 over the last two decades has demonstrated a steady increase that is almost certainly man-caused (Ekdahl and Keeling 1973). The increase in atmospheric carbon has been attributed primarily to the burning of fossil fuels (Woodwell and Pecan 1973; Baes et al. 1976, 1977) although some investigators have hypothesized that worldwide deforestation may also represent a net source of CO_2 to the atmosphere (Bolin 1977, Woodwell et al. 1978, Adams et al. 1977). Probably the most serious implication of the increase in CO_2 concentration could be a global warming caused by the "greenhouse effect" of atmospheric CO_2 (Baes et al. 1977).

Potential ramifications of the full-scale use of biomass for energy for the global carbon cycle merit consideration. Section 5.5 discusses marine biomass production and the global carbon cycle. Three major points in the consideration of terrestrial biomass for energy warrant discussion: (1) the CO_2 released from combustion of biomass or biomass-derived fuels is nonfossil carbon, which would eventually cycle to the atmosphere as part of the natural carbon cycle; (2) the land biota could represent a CO_2 sink for removal of some excess atmospheric CO_2; and (3) the carbon reservoir of humus and dead organic matter in the soil is greater than that in the land biota and could serve as a source of atmospheric CO_2 in the event of increased conversion of forests to managed agricultural systems.

The carbon contained in biomass and biomass-derived fuels would eventually cycle to the atmosphere whether or not it is used to produce energy for humans. Energy from biomass would thus represent some decrease in the amount of fossil CO_2 that enters the global carbon cycle. However, the decrease is likely to be relatively minor because energy from biomass will supplant only a small percentage of current fossil fuel use. Under the "maximum practicable" case analyzed in the domestic policy review of solar energy (Solar Energy Task Force 1978), biomass may be expected to provide about 6×10^{18} J (6 quads) of energy by the year 2000, which is less than 8% of the current U.S. energy use of $\sim 78 \times 10^{18}$ J (74 quads). The more realistic "base" case predicts about 3×10^{18} J (3 quads) by the year 2000, less than 4% of current U.S. energy demand (Solar Energy Task Force 1978). Furthermore, the current estimate of world net primary productivity from the terrestrial biosphere of about 53×10^{15} g of carbon per year (Woodwell et al. 1978) is only an order of magnitude greater than the current total annual input of 5×10^{15} g of carbon from fossil fuel burning (Baes et al. 1977). It is clear that biomass energy will never supply a major portion of the energy now derived from fossil fuels. The incremental decrease in fossil CO_2 input from the replacement of a small fraction of fossil fuels by biomass fuels is not likely to affect significantly the global carbon budget.

A widespread use of biomass for energy is of concern because it involves a major component of the global carbon cycle, the terrestrial biota. The atmospheric concentration of CO_2 has increased by only half as much as the total amount of CO_2 released from fossil fuel burning (Baes et al. 1977). About one-half of the excess atmospheric fossil CO_2 has been taken up by one or both of the other major components of the global carbon cycle—the oceans and the terrestrial biosphere. Considerable controversy has occurred recently over whether the land biota has served as a sink or as a source of atmospheric CO_2 over the last century (Woodwell and Pecan 1973, Botkin 1977, Woodwell et al. 1978). However, the hypothesis that accelerated worldwide clearing of primary forests, mainly in the tropics, has represented a net source of atmospheric CO_2 over the last few decades has been generally accepted (Bolin 1977, Wong 1978, Adams et al. 1977, Woodwell et al. 1978). If the increasing emphasis on biomass for energy causes a significant reduction in the acreage of old climax forests in the United States, a short-term increase in atmospheric CO_2 could result. Such is not expected to be the case, however, because forest lands harvested for energy in the United States would be promptly reforested, probably as silvicultural energy farms (see Chap. 5). Thus the input of CO_2 to the atmosphere from burning the forest biomass for energy would be more than balanced by the CO_2 fixation in the rapidly photosynthesizing new forest growth.

A third mechanism whereby use of biomass for energy could affect the global carbon cycle concerns the carbon reservoir in the soil. The carbon reservoir in humus and dead organic matter in forest soils is estimated to be from 1.5 to 4 times greater than that in living land biota (Woodwell et al. 1978, Bolin 1977). Furthermore, the turnover rate for the soil carbon is generally longer than that for the land biota (Bolin 1977). Conversion of wetlands and forested lands to managed agricultural ecosystems results in a release of carbon from the soil by accelerated decay of soil organic matter (Baes et al. 1976, Woodwell et al. 1978, Bolin 1977). Thus the conversion of marginal forests to herbaceous energy farms would definitely result in a short-term net increase in CO_2 input to the atmosphere. Further research concerning nutrient cycling and whole-tree harvesting is necessary to determine if such would also be the case for conversion of forested land to silvicultural energy farms.

In conclusion, the decrease in use of fossil fuels resulting from displacement by biomass fuels will probably not be sufficient to affect significantly the input of fossil CO_2 to the atmosphere. Any incremental decrease in fossil CO_2 to the atmosphere could be more than offset by release of CO_2 from conversion of forests to managed agricultural ecosystems for the production of biomass energy, particularly if poor soil management practices are employed (see Chap. 5). However, the preceding discussions of biomass for energy and the global carbon cycle are at best very preliminary generalizations from the recent literature on the carbon cycle. Considerably more research is necessary to make any definite predictions. In particular, the estimates of standing terrestrial biomass, world net primary productivity, and changes in the carbon flux between the terrestrial biosphere and the atmosphere need to be refined.

The extent and mechanisms for carbon flux between the oceans and the atmosphere need investigation, especially the carbonate buffering system of the open ocean, because the ocean appears to be a greater sink for CO_2 than was recently estimated (Wong 1978, Stuiver 1978). Estimates of the future rate of fossil CO_2 release must be updated in view of the various alternative energy systems that are being developed. Finally, the potential climatic ramifications of atmospheric CO_2 need to be better understood.

References

Adams, J. A. S., M. S. M. Mantovani, and L. L. Lundell, 1977. "Wood Versus Fossil Fuel as a Source of Excess Carbon Dioxide in the Atmosphere: A Preliminary Report," *Science* **196:** 54–56.

Baes, C. F., Jr., et al., 1976. *The Global Carbon Dioxide Problem,* ORNL-5194, Oak Ridge National Laboratory, Oak Ridge, Tenn.

———, 1977. "Carbon Dioxide and Climate: The Uncontrolled Experiment," *Am. Sci.* **65**(3): 310–20.

Benemann, J. R., 1978. *Biofuels: A Survey,* ER-746-SR, Electric Power Research Institute, Palo Alto, Calif.

Bolin, B., 1977. "Changes of Land Biota and Their Importance for the Carbon Cycle," *Science* **196:** 613.

Botkin, D. B., 1977. "Forests, Lakes and the Anthropogenic Production of Carbon Dioxide," *Bioscience* **27**(5): 325–31.

Braunstein, H. M., et al., 1978. *Environmental and Health Aspects of Solid Wastes from Coal Conversion: An Information Assessment,* ORNL-5361, Oak Ridge National Laboratory, Oak Ridge, Tenn.

Calef, C. E., 1976, "Not Out of the Woods," *Environment* **18**(7): 17–25.

Capar, S. G., et al., 1978. "Multielement Analysis of Animal Feed, Animal Wastes, and Sewage Sludge," *Environ. Sci. Technol.* **12:** 785–90.

Cramer, O. P., 1974. "Air Quality, Fire-Hazard, and Esthetics," pp. F1–F51 in *Environmental Effects of Forest Residues Management in the Pacific Northwest—A State-of-Knowledge Compendium,* O. P. Cramer, ed., For. Serv. Gen. Tech. Rep. PNW-24, U.S. Department of Agriculture, Washington, D.C.

DiNovo, S. T., et al., 1978. *Preliminary Environmental Assessment of Biomass Conversion to Synthetic Fuels,* EPA-600/7-78-204, Industrial Environmental Research Laboratory, Cincinnati, Ohio.

Dreisbach, R. H., 1974. *Handbook of Poisoning,* 8th ed., Lange Medical Publications, Los Altos, Calif.

Ekdahl, C. A., Jr., and C. D. Keeling, 1973. "Quantitative Deductions from the Records at Mauna Loa Observatory and at the South Pole," in *Carbon and the Biosphere,* G. M. Woodwell and E. V. Pecan, eds., Atomic Energy Commission, Technical Information Center, Washington, D.C.

Electric Power Research Institute, 1978. "Executive Summary," Workshop on Biomass Energy Technology, Santa Clara, Calif., Nov. 8–9, 1978, EPRI AF-1036-SY, Electric Power Research Institute, Palo Alto, Calif.

Fraser, H. S., and E. P. Swan, 1972. *Chemical Analyses of Veneer-Dryer Condensates,* Rep. VP-X-101, Western Forest Products Laboratory, Vancouver, British Columbia, Department of Environment, Canadian Forestry Service.

Gikis, B. J., et al., 1978. *Preliminary Environmental Assessment of Energy Conversion Processes for Agricultural and Forest Product Residues,* EPA-600/7-78-047, NTIS PB-281 189, Municipal Environmental Research Laboratory, Cincinnati, Ohio.

Golembiewski, M., 1978. *Environmental Assessment of a Waste-to-Energy Process Wood and Oil Fired Power Boiler,* draft final report prepared for the Environmental Protection Agency by Midwest Research Institute, Industrial Environmental Research Laboratory, Cincinnati, Ohio.

Hodam, R. H., and R. O. Williams, 1978. "Small Scale Gasification of Biomass to Produce a Low Btu Gas," pp. 729–48 in *Proc. of Conf. on Energy from Biomass and Wastes, Aug. 14–18, 1978, Washington, D.C.,* Institute of Gas Technology, Chicago, Ill.

Jahnsen, V. J., 1961. *The Geochemical Composition of Hardwood Smoke,* Ph.D thesis, Purdue University, Lafayette, Ind.

King, N. J., R. W. Mungomery, and C. G. Huges, 1965. *Manual of Cane Growing,* American Elsevier Publishing Co., New York.

Milliken, J. O., 1979. "Airborne Emissions from Wood Combustion," presented to the Wood Energy Institute Wood Heating Seminar IV, Mar. 22–24, Portland, Ore.

National Academy of Sciences, 1972. *Particulate Polycyclic Organic Matter,* Committee on Biologic Effects of Atmospheric Pollutants, Washington, D.C.

———, 1977*a*. *Carbon Monoxide,* Committee on Biological Effects of Atmospheric Pollution, Washington, D.C.

———, 1977*b*. *Nitrogen Oxides.* Committee on Medical and Biologic Effects of Environmental Pollutants.

Roop, R. D., 1978. "Energy from Biomass: An Overview of Environmental Aspects," pp. 33–42 in *Proc. 2nd Nat. Conf. on Technology for Energy Conservation, Jan. 24–26, Albuquerque, N.M.*

Ruch, R. R., H. J. Gouskater, and N. F. Shimp, 1974. *Occurrence and Distribution of Potentially Volatile Trace Elements in Coal: A Final Report,* Illinois State Geological Survey, Environ. Geol. Note 72, Urbana.

Shelton, J. W., 1976. *The Woodburners Encyclopedia,* Vermont Crossroads Press, Waitsfield, Vt.

Solar Energy Task Force, 1978. *Solar Energy Domestic Policy Review, Status Report,* prepublication public review copy.

Stuiver, M., 1978. "Atmospheric Changes and Carbon Reservoir Changes," *Science* **199:** 253–58.

Thornton, T., 1978. *An Assessment of Anaerobic Digestion in U.S. Agriculture,* ESCS-06, Economics, Statistics and Cooperatives Service, U.S. Department of Agriculture, Washington, D.C.

Turner, D. B., 1970. *Workbook of Atmospheric Dispersion Estimates,* EPA Report AP-26, U.S. Environmental Protection Agency, Research Triangle Park, N.C.

U.S. Environmental Protection Agency, 1977. *Compilation of Air Pollutant Emission Factors,* 3d ed., EPA Rep. AP-42, NTIS PB 275 525, Office of Air and Waste Management, Research Triangle Park, N.C.

———, 1978*a*. *Electric Utility Steam Generating Units—Background Information for Proposed NO_x Emission Standards,* EPA/450/2-78/005A.

———, 1978*b*. *Preliminary Environmental Assessment of Biomass Conversion to Synthetic Fuels,* EPA-600/7-78-204.

Varani, F. T., J. Burford, and R. P. Arber, 1977. "The Design of a Large-Scale Manure/Methane Facility," in *Fuels and Energy from Renewable Resources,* D. A. Tillman, J. U. Sarkanen, and L. L. Anderson, eds., Academic Press, New York.

Wong, C. S., 1978. "Atmospheric Input of Carbon Dioxide from Burning Wood," *Science* **200:** 197–200.

Woodwell, G. M., and E. V. Pecan, eds., 1973. *Carbon and the Biosphere, Proc. of the 24th Brookhaven Symposium in Biology, May*

16-18 1972, Upton, N.Y., U.S. Atomic Energy Commission, Technical Information Center, Washington, D.C.
Woodwell, G. M., et al., 1978. "The Biota and the World Carbon Budget," *Science* **199:** 141-46.
Yang, V., and S. C. Trinidade, 1978, "The Brasilian Gasohol Program," in *Energy from Biomass and Wastes, Proc. of the IGT Symposium, Washington, D.C., Aug. 14-18, 1978,* Institute of Gas Technology, Chicago, Ill.

7

Social and economic implications of large-scale production of energy from biomass

To weigh adequately the pros and cons of biomass energy production in any of its diverse configurations, both the products and the requirements of specific biomass technologies must be expressed in human terms—a process that involves direct examination of aesthetics; employment and income impacts; population growth and settlement patterns; political, social, and economic interactions; etc. This chapter provides a brief review of efforts that have already been made in this direction and attempts to highlight both the issues now considered to be of major importance and areas that appear to be most in need of further research. In the course of this review, two key findings have become clear:

- The large-scale development of biomass as a renewable source of energy involves social and economic issues of critical importance. These issues can be expected to constrain significantly both the rate of development and the ultimate potential of biomass as an energy source.
- Major gaps exist in our understanding of the probable social and economic impacts of most large-scale biomass energy production schemes. Filling these gaps will require a major commitment in time and other resources.

Biomass differs from conventional energy technologies in its diffuseness and (relatively speaking) extremely large land requirements, in the scale and potential pervasiveness of production and conversion facilities in rural settings, in its juxtapositioning of energy with a host of previously unrelated activity sectors, and in other respects. Although analysis and extrapolation from previous energy and biomass production (e.g., forestry and agricultural) experience will be useful, it is clear that fertile imaginations will need to be brought to bear in projecting the social and economic implications of large-scale biomass energy production.

This review is designed to identify both a set of major generic socioeconomic issues and a sampling of the more resource- or technology-specific impacts associated with large-scale biomass development. Eight areas of "macroscopic" interest—national and international issues and/or impacts that appear to be associated with several different biomass options—are covered in Sect. 7.1, and an effort to cover the most important aspects primarily of near-term applications occupies most of the remainder of the chapter (i.e., Sect. 7.2). Specific discussions in the latter case have been hampered by a lack of prior impact analysis work. Because the commitment necessary to fill existing gaps in information is well beyond the scope of this review, a focus has been developed in the treatment of each near-term application: the discussion of wood emphasizes a centralized (50-MW direct combustion) conversion plant, and new producer requirements and changes in farming practice are highlighted with regard to agricultural biomass; methane production raises a set of

issues perhaps unique to a combined production/conversion operation, and discussions of ethanol emphasize market development and the implications of large-scale subsidies.

The presentation that follows should therefore not be considered exhaustive. In fact, major issues may well have been overlooked. Instead, it is intended to provide an overview of the type of issues that may be encountered while making the best use of information available. Readers are encouraged to review the chapter in its entirety, even if they are only interested in a particular application or level of impact, and to develop specific first-order questions in their area of interest by analogy (e.g., comparing centralized distilleries with wood-fired power plants) or by extension (e.g., what institutional issues would arise in conjunction with freshwater biomass production?). Analogy and extrapolation may be of very limited usefulness in terms of specific impacts, but these techniques should help to provide an initial set of questions as a starting point for technologies that have not yet been investigated in detail.

Existing assessments of the possible socioeconomic impacts of large-scale energy production from biomass suggest that such impacts are likely to be both pervasive and diverse. Although biomass facilities are generally smaller than conventional coal or nuclear power plants, photosynthesis, and hence production and conversion of biomass, can take place in almost any location. Impacts can be expected at the individual, corporate, community, regional, national, and international levels, and virtually all sectors of social and economic activity may ultimately be affected. In particular, farms, ranches, forest products operations, and existing utility and fuel production interests will certainly be affected directly. Additional direct impacts are likely in food, construction, transportation, farm machinery, fertilizer, and other industries. Countless indirect impacts ranging from effects on local retail trade, housing stocks, opportunities for recreation, and the demands placed on community services to changes in future tax revenues, regional employment, and world food supplies must also be anticipated. We hope that the brief review presented here will help to stimulate further research efforts to provide a broadening and timely knowledge of the full implications of biomass development.

7.1 Generic Impacts and Issues

Eight major areas of generic interest are considered in this section: land use, market linkages and instabilities, the potential for overexpansion in biomass energy production efforts, long-term food supply implications, employment and demographic changes, institutional and political issues, and transportation. The primary concerns in these areas are summarized in Table 7.1.

7.1.1 Land Use

Because of the low (about 1%) efficiency of photosynthetic energy conversion, large-scale biomass energy production will require millions of acres of land—probably more land per unit of energy produced than any other energy technology. For comparison, western coal sufficient to produce 0.021 EJ ($\sim$0.02 quad) can be mined annually with an average disturbance of only about 100 ha (250 acres), assuming reclamation after five years; for thinner seams in the East and Midwest, about 607 ha (1500 acres) would be disturbed for similar yields (Reuyl et al. 1977). Comparable fuel production from biomass energy plantations would, however, require from 60,730 ha (150,000 acres) to more than 202,430 ha (500,000 acres) (Sect. 3.1.4). In other words, biomass land requirements are two or three orders of magnitude larger than those for coal conversion per unit of energy produced. They are between one and two orders of magnitude larger than those of direct solar conversion per unit of energy produced.

Although hundreds of millions of hectares of "underutilized" agricultural, range, and forest lands may be available for biomass energy production, much of this land is seriously limited in terms of cultivated crop production. Major portions are either located in arid or semiarid parts of the country (McElroy 1977) or held in relatively small or widely distributed parcels by a large number of independent owners (Alich et al. 1977, Bentley 1979).

Table 7.1. Generic socioeconomic impacts of biomass energy systems

Area of interest	Impact or issue description
Land use	Land requirements for biomass energy production are orders of magnitude larger than for other energy technologies. This raises significant aesthetic and practical impact and land-availability questions. Severe conflicts among competing land-use interests can be expected.
Market linkage	Linkages among biomass product markets can be complex and can lead to either increasing or decreasing overall market stability. Diversity in feedstock and product mix and vertical or horizontal process integration should probably be encouraged.
Dynamics of market expansion	Excessive subsidization, stored resources, and probable expansion of more-profitable biomass markets could imply wasteful and destructive overexpansion of biomass energy production.
Food	Competition between food and energy from biomass could affect domestic and international food supplies and prices.
Employment and demographic changes	Location of biomass facilities in rural areas would generally produce significant employment and community impacts. Regional demographic shifts may also occur as a result of variations in biomass availability.
Institutional issues	Relationships among biomass energy producers, existing utilities, and potential customers for cogenerated steam need to be developed. Environmental regulations, tax treatment and government revenue implications, and capital availability are also major areas of uncertainty.
Political issues	Decentralization and subsidies among competing energy programs may become even more prominent political issues.
Transportation	Road upgrading and maintenance requirements as well as traffic pattern changes may be significant. Indirect impacts of transportation system alterations should also be considered.

Although the most desirable land reservoirs [for example, 30 to 40 million ha (75 to 100 million acres) of cropland not in crops] could well be shifted to biomass production, such a shift might also impair crop production capabilities (McElroy 1977). Conflicting land development interests also have an impact on land availability for biomass; "...during the past 20 years some 11 million ha [of cropland, pastures, and forests] (an area larger than Ohio) have been converted to urban areas and highways" (Pimental et al. 1975). Major conflicts are also likely between biomass production and recreational use undeveloped lands. On balance, severe conflicts with competing land-use interests appear to be inevitable and may greatly restrict the development of biomass as an energy source. (See Sect. 5.2.1 for a more detailed treatment of land availability and for a discussion of erosion and other potential biomass land-use impacts.)

Direct public resistance to extensive land- or water-based biomass development may become a highly significant or binding constraint in itself. Large groups of people feel aesthetic and other impacts caused by both the amount of land required and the distribution of available plots. Charles F. Bennett, of the University of California at Los Angeles, suggests with reference to forest resources that "... even with such ... traditionally sanctioned [uses as] timber harvest, emotions [can] wax very hot and cause great difficulties for the Forest Service." "The public perception of National Forests being harvested on a scale large enough to meet any appreciable part of this nation's demands for Btu's would be, I believe, an apocalyptic vision and one so pervasive as to rule out for many decades any possibility of such use being made of 'natural' forests in the public domain" (Bennett 1979). Extensive development of *any* "underutilized" and hence relatively natural areas for energy purposes could encounter similar perceptions (see Sect. 7.4).

7.1.2 Market Stability and Balance of Trade Benefits

The economic, as well as the net, energy productivity of a biomass energy system often depends on the full and complete utilization of a variety of products from the biomass resource (Lipinsky 1978, Edminster 1976). Herendeen (1978) argues, for example, that ethanol production from corn becomes a net energy producer only when the assumption is that stalks and cobs can be marketed or used directly as a boiler fuel. The number of products made directly from a single biomass feedstock can be surprising: corn cobs are used to produce furfural, Slikwik (an absorbent used in cleaning up oil-tanker spills), cattle feed, absorbent carriers for agrichemicals (e.g., to provide time release of pesticides), industrial abrasives, and xylatol (a new sweetener). Cobs can also be burned directly or left in the field as a soil conditioner. The linkage of biomass products with biomass product-related markets can be quite complex and will require careful consideration in some cases (Hudson 1977, McAllister 1976).

Complicated relationships among product markets can lead to greater stability in both individual and overall markets as the price and supply variations in one area are smoothed through changes in the biomass product mix. Profitable by-product industries (energy production is often one of these) that might otherwise be highly vulnerable may be sustained by larger, more stable industries that bear major portions of the biomass production, collection, or conversion costs. For example, energy from biomass has been widely acknowledged to permit the sugar industry in Hawaii to survive in the face of depressed sugar prices and escalating petroleum costs (U.S. Department of Energy 1978).

Market linkages can, however, lead to problems. A publicly owned company in Ireland, for example, began converting surplus potatoes into alcohol in the late 1930s with the explicit hope of stabilizing the potato market. Because supplies were difficult to obtain except in times of glut, most of the instability was simply transferred into the operation of the industrial company concerned. Government regulations were ultimately necessary to force gasoline distributors to purchase the alcohol that was produced (McAllister 1976).

Where none of a set of biomass-related markets is independently viable, instabilities in one area may be transmitted to other areas in ways that amplify instabilities (McAllister 1976). If marginal cropland were developed based on the expectations of returns from both food and energy products, adverse variations in either market might lead to abandonment and subsequent impacts on the other related product market. Even when a project is not abandoned, instabilities in one or another submarket could transform it from an initially energy-producing to a net energy-consuming enterprise. Independently viable food and energy markets will also feel the impacts of inevitable variations in weather and other natural and market conditions. Periodically idle capacity may be unavoidable in some cases.

The regional and national balance-of-payments benefits of displacing imported fuels may be an important benefit of biomass energy production. Such benefits can only be expected, however, when a biomass energy system is, in fact, an energy producer and one that does not significantly rely on continuing subsidies to make up the difference between real (subject to market variation) and theoretical (perfect utilization) operations. As a potentially important exception to this rule, balance-of-payments benefits could also be achieved by using biomass systems (in effect) to convert domestic solid fuels into liquid fuels (e.g., coal used as the source of heat for distillation of ethanol) to displace imported oil (Herendeen 1978).

To minimize the adverse impacts of instability, it is recommended that diversity be encouraged (Alich 1977, Black 1977, and Abeles and King 1978) in both the product and the feedstock segment of a biomass enterprise. Combinations of various feedstocks can improve process efficiency (e.g., combinations of sawdust and manure for methane production) as well as substantially improve the economics of a plant by providing for year-round operation without large biomass storage requirements (Alich 1977). Vertical or horizontal integration of one or more stages in the production, processing, and marketing complex could also help to ensure adequate feedstock supplies and continued economic viability as product prices vary (Alich 1977, Abeles and King 1978, and Black 1977).

7.1.3 Dynamics of Market Expansion

With reference to liquidation of excess inventories of public timber, Olson (1977) mentions the problem of "falldown" in timber supply following conversion of "excess" forest resources. Several factors, including major subsidization, an initially large physical potential, and a relatively sudden economic emergence, appear to contribute to the overextension of many biomass energy alternatives. Storage of biomass resources, both directly (e.g., in the form of unharvested forests) and indirectly (e.g., in the form of existing but easily depleted soil and nutrient resources), may present a particular problem.

Future growth in existing, more profitable biomass product markets, such as paper manufacture, food and lumber production, etc., must also be taken into account (Alich 1977). With rising petroleum costs and other developments, plastics, lubricants, detergents, and other new products could also begin to compete for biomass materials (Edminster 1976). Growth of this sort in old and new product markets could result in a kind of "retroactive overextension" of biomass energy production as feedstocks become diverted to potentially more profitable applications. Simple mobilization of biomass resources as one element of any energy production system could in itself add to active competition with energy conversion facilities as new markets develop for previously unavailable feedstock (Lipinsky 1978).

Overextended development of biomass for energy could lead to accelerated depletion of marginal and even productive land as marginally profitable concerns shift from long-term sustainable yield management to an interest in immediate economic survival.

7.1.4 Food

The long-term relationships between energy and food production from biomass will require careful consideration. A 37.9 million m^3/year (10 billion gal/year) grain ethanol program, for example, would require substantial increases in grain production to satisfy both food and energy needs and acreage requirements which "would result in such dramatic changes in crop mix and ... land rents that the impacts on other crops cannot easily be estimated" (U.S. Department of Agriculture 1978). As another example, if all our gasoline were to be replaced by gasohol containing 10% ethanol and all that ethanol were to be obtained from corn (an illustrative, not a likely, prospect), the equivalent of all our corn exports and 73% of the corn fed to livestock (a total of 110 million metric tons) would be required (Chambers et al. 1979 and U.S. Department of Agriculture 1978). Competition for land and water resources will not be the only issue, because fertilizer prices, transportation costs, the demand for investment resources, and many other factors clearly link food and biomass energy markets (McAllister 1976).

Because the major source of foreign revenue for the United States is its export of agricultural commodities (Hunter et al. 1976), competition between energy and food production could seriously detract from any reduction in oil trade deficits. In addition, domestic and international food prices could be affected by direct supply variations (e.g., diversion of sugarcane to alcohol production), by indirect product availability (e.g., effects on beef availability caused by diversion of corn from cattle feed to ethanol production), and by indirect shifts in crop mix (e.g., additional wheat for alcohol, displacing soybeans).

Food production can no longer be increased significantly by mobilizing new tracts of arable land in the United States (Pimentel et al. 1976). Arable land is, in fact, already insufficient to feed the world's population a diet similar to that consumed in the United States because of the caloric and protein inefficiency of meat consumption (Pimentel et al. 1975). Biomass energy technologies must be considered carefully and individually in the light of heavy current and future world dependence on U.S. food production. Questions such as "Is it right to take food from the world's hungry and put it in America's automobiles?" may well have to be addressed (Anderson 1978).

In fairness, it appears that combined biomass and fuel production would have a beneficial effect on farm incomes, at least in the short run. Such operations would also increase resource utilization, reduce existing excess capacity, and increase overall marketable production. Consideration needs to be given to how extensive and long lived these benefits are likely to be and to whether or not they outweigh possible detrimental impacts.

7.1.5 Employment and Demographic Changes

In general, biomass energy systems appear to be more labor intensive than their fossil fuel counterparts (DiNovo et al. 1978). Employment effects vary, however, depending on the particular technologies under consideration. Methane production from small farm feedlots (25 sows, 300 steers), for instance, would not necessitate direct additions to the typical farmer plus full-time hired man labor force (Iowa State University 1977); extensive additional harvest and utilization of wood, on the other hand, could be a direct and significant source of new jobs (Groncki et al. 1978). Where a net increase in jobs is likely, consideration should be given to the availability of local workers (e.g., people operating marginally productive farms or small businesses), to the extent to which outsiders would have to be brought in, and to the type of jobs each group might fill.

Because of the high cost of biomass handling and transportation, it is expected that economically viable processing facilities will have to be located near biomass growth and/or collection areas (DiNovo et al. 1978 and Alich and Inman 1976). Conversion, as well as growth and collection activities, will therefore be centered to a large extent in rural areas and will produce potentially major impacts on small towns. Housing, water supply, fire protection, transportation, school, police, medical, and other services and facilities will in some cases be inadequate to handle directly and indirectly induced population growth (Samuels 1979 and Hewett and High 1978*a*). Income distributions may be altered (DiNovo et al. 1978 and VanderWerf 1978), and significant adjustments may be involved in shifting from the construction to the operating phase of a conversion facility (VanderWerf 1978). The socioeconomic implications of locating large conversion plants in rural environments is a research area that needs a great deal more attention (DiNovo et al. 1978).

To the extent that bioconversion operations develop near urban centers, socioeconomic changes may prove relatively minor (DiNovo et al. 1978). In both urban and rural situations, "... physical and socioeconomic impacts [will be somewhat] distributed, being smaller in any one community although more communities may be affected" than with larger conventional energy conversion facilities. Costs and benefits may be relatively localized, with both new employment and increased public service costs focused in the same area (Hewett and High 1978*a*). Regional demographic shifts, such as those associated with industrial plant location (Samuels 1979) or regional balance of trade, could be significant for major development of region-specific biomass resources.

7.1.6 Institutional Issues

At least three institutional issues are likely to require careful consideration in the development of biomass energy sources: (1) interaction with existing electric or gas utilities (where gas or electricity is to be produced), (2) tax treatment and associated government revenue effects, and (3) capital availability for, and financing of, biomass facilities.

With the development of onsite electricity production from wood, methane, or other biomass resources, rate structures for excess power sales to the utility grid and for backup power purchases from the grid will need to be negotiated and cleared through state public utility commissions. The regulatory status and standards for grid connection of power or gas producers will also need to be resolved (Alich et al. 1977). Under what conditions, for example, should a gas supplier be expected to connect his major pipeline to, and purchase gas from, a methane-producing feedlot?

The time required to construct a biomass conversion facility will generally be significantly less than that required for a large-scale conventional plant—a fact that may make such facilities far more attractive to existing utilities (Blake and Salo 1977; *Energy Insider*, Apr. 2, 1979). It may be of interest to note that to escape the complications of federal and state regulations for an independent public utility, an independent organization interested in grid-connected cogeneration may be forced to offer the existing electric utility the opportunity to construct and operate the generating facility (Berger 1977). This complication may sometimes act as a deterrent in the planning and financing of such facilities.

Existing tax laws may require review or revision to ensure equitable treatment of new biomass energy industries (Segerson 1978). In addition, if tax incentives such as the current Nebraska 5¢/gal (Anderson 1978) and federal 4¢/gal (Bentley 1979) gasoline tax exemption for gasohol (10% alcohol) are to be provided, new taxes may be necessary to make up for lost revenues.

Biomass harvest and conversion activities involve significant capital investment in facilities and equipment. In some cases, total revenues may be relatively small compared with the necessary investment, and it may be difficult to attract capital (Alich et al. 1977). Many existing industrial concerns such as sawmills or pulp mills may be too small to afford biomass conversion systems that would be compatible with their operations, and they could lose competitive standing relative to their larger counterparts as a result. Cooperative ventures (e.g., farmers' cooperatives) have been suggested as a means of providing for both local ownership and stabilized feedstock supplies (Alich et al. 1977 and Black 1977).

Two other institutional interface issues should be mentioned: (1) environmental, health and safety, and other government regulations, which generally do not yet specifically address biomass energy systems, may have a significant impact on future paths of development; and (2) where biomass is used to fuel a cogeneration system, steam may be sold to a third party (someone other than the electric utility). Cost allocations among energy products and other aspects of this interaction may present significant issues in themselves (Hewett and High 1978*a*).

7.1.7 Political Issues

Decentralization of energy supply systems could become more of an issue with the development of biomass. The economics of transporting biomass somewhat reduces normal conversion plant economies of scale; small, local, and perhaps cooperatively owned plant concepts may therefore compete with larger, more centralized concepts in some cases. Peters (1979) suggests the possibility of very small distilleries, located only 40 to 50 miles apart in producing areas, designed to produce 170- to 180-proof ethanol for direct local use as an undiluted fuel. This approach would require less energy for distillation [gasohol must be made with dry (near 200-proof) ethanol]. In recommending further study of decentralized distilleries before adoption of the large-scale designs characteristic of Nebraska, a Minnesota state legislature working group also suggested that it might be possible to feed the distiller's dried grains (a distillery by-product) to cattle more promptly with decentralized facilities and thus save the energy otherwise required for drying the feed (Abeles and King 1978). To the extent that these and other decentralized concepts compete with centralized systems, political considerations may become significant.

Political friction is clearly to be expected as biomass regions and other interest groups press for nationally subsidized development programs in competition with other energy interests. Some oil companies "have been accused of launching a 'cunning nationwide campaign' against gasohol and of pouring out 'clumsy distortions of the truth' " (Anderson 1978). This sort of friction is exacerbated by the variety of often equally defensible assumptions that can be adopted in calculating the energy or economic returns associated with biomass and other energy technologies.

It must be borne in mind that the "stable and often homogeneous areas likely to be impacted [by biomass] development may prefer to forego [its] benefits altogether. Public sentiment may determine that maintenance of the status quo best represents local interests." At best, such sentiments are likely to contribute to local political friction in that a perfect consensus seems improbable in most cases (DiNovo et al. 1978).

7.1.8 Transportation

"Costs of transport vary considerably depending on four factors: load density, regularity, load and unload conditions, and distance" (Miles 1977). In many cases, transport costs are the largest contributors to the overall cost of biomass energy production (Blake and Salo 1977), whereas in other cases transport costs are relatively small (Tinberg and McElroy 1978). Because of the bulk, wide distribution, and high water content of most raw biomass resources, however, the dollar, energy, and social costs of transportation should receive careful consideration in sizing and in siting biomass facilities; in many cases it may be most efficient to utilize energy produced from biomass at the growth and harvesting site (Vergara and Pimentel 1978). Transport costs (costs per kilogram-kilometer) decline with distance (Blake and Salo 1977), but most of that decline may occur in the first 40 km (25 miles). [Tinberg and McElroy (1978) project costs for grains and grasses at 0.016¢/kg-km for 8 km, 0.0089¢/kg-km for 40 km, and 0.0079¢/kg-km for 80 km.]

Although rail transport may be appropriate in some cases, trucks generally appear to be the most convenient and least expensive means now available for moving biomass (Inman 1977, Adler et al. 1978, Blake and Salo 1977, and Barbera 1979). Truck noise, traffic, and road upgrading and maintenance effects need to be taken into consideration, however, because they may be significant for more centralized (e.g., 50-MW wood-fired power plants) conversion facilities (Adler et al. 1978). Trucking impacts will depend in part on who owns and operates the trucks. "It is widely acknowledged that private truckers frequently exceed legal weights"—a practice that accelerates road surface deterioration and increases the risk of bridge collapse in rural areas (Adler et al. 1978). As another example, private harvest and use of wood resources for home heating (wood stoves) would tend to result in a larger number of smaller loads than would be the case with dedicated trucking to a wood-fired power plant. Diesel fuel sales and other secondary impacts can also be significant (see Sect. 7.2.1).

Additional transportation requirements may arise due to regional imbalances in biomass energy supplies and local energy demand. In Sao Paulo, Brazil, ethanol production apparently already exceeds state requirements for a 20% gasohol blend, and the excess may have to be trucked to neighboring states (Yang and Trindade 1978).

The transportation requirements interact strongly with other aspects of biomass utilization systems, including harvest, storage, and conversion. Densification and modularization for easier handling, as well as drying and other procedures, have been areas of some interest and experimentation (Miles 1977). In addition, handling of transportation systems may lead to a variety of secondary effects, including either increased suburbanization of rural areas and cultural homogenization (e.g., with improved road systems) or growing disparities between rural and urban cultures (e.g., with poorly maintained roads and localized use of biomass resources).

Further treatment of transportation issues is included in other sections of this chapter, including Sects. 7.2.1 and 7.2.2.

7.2 Potential Impacts of Specific Near-Term Applications

The discussion that follows is designed to provide a more detailed view of the potential socioeconomic impacts of near-term biomass technologies. The details presented here are *not* to be taken as an exhaustive or even necessarily a representative review. Instead, they are intended as a "sampling" and as an indication of the uniqueness and variety of system location, scale, and design impacts that will require consideration. As stated earlier, a focus has been developed for each near-term application reviewed here: the discussion of wood emphasizes a centralized (50-MW direct combustion) conversion plant, whereas new producer requirements and changes in farming practice are highlighted with regard to agricultural biomass. A set of problems perhaps unique to combined production/conversion operation is outlined in the section on methane production, and discussions of ethanol emphasize market development and the implications of large-scale subsidies. Although some discussion of matters outside these areas of concentration is provided for each biomass alternative, it should be clear that major gaps remain to be filled. The reader is again encouraged to review the section in its entirety and to formulate questions for further investigation in particular areas of interest by drawing analogies among the technologies.

7.2.1 Forest Energy Production and Centralized Conversion

A fairly detailed study of the potential for small wood-fired power plants in northern Vermont has recently been conducted at Dartmouth's Thayer School. The employment and income impacts of plant construction and operation, as well as issues such as the availability, harvest, and transportation of wood resources, have been analyzed for a nominal 50-MW electric power plant. Perhaps because of their demonstrated commercial viability in certain areas (U.S. Department of Energy 1978 and Maullin et al. 1979), direct-combustion power plants have received more detailed attention in terms of socioeconomic impacts than many other forest biomass technologies; the Dartmouth study results are reviewed here as an introduction to some of the specific impacts that may arise in conjunction with forest energy production and with centralized conversion facilities located in rural settings.

The most obvious requirement for a wood-fired power plant (WFPP) is that an adequate supply of wood be readily available. Hewett (1978) indicates that 92% of Vermont's commercial forest land is privately owned and that interest in timber harvesting tends to be related directly to the size of the land holdings. In addition, he has found that noncorporate, private landowners who will consider removal of low-quality forest material often will not permit whole-tree harvesting. The "economics of mechanized harvesting can be shown to discriminate against small holdings" as well, and Hewett assumes that only about one-third of Vermont's private forest land will be available for whole-tree chipping of low-quality trees. The proximity of these resources to roads and potential WFPP sites will, of course, further limit fuel availability as will price increases, which are likely to be "significantly faster" than inflation. The inability of independent contractors to raise capital for mechanized harvesting equipment and the lack of organizational skills necessary for high-productivity operations are likely to slow the growth of wood-chip supplies and aggravate price increases caused by growing reliance on chips for construction materials, furniture stock, paper, and fuel.

In spite of these restrictions, five towns in northern Vermont appear to have "good potential" as WFPP sites on the basis of an initial screening performed by High (1978).In addition to wood availability, High considered the following initial screening criteria: (1) plant site availability (i.e., 100 acres of unused, nearly level land not subject to flooding, etc.); (2) transportation networks (roads or railroads either adequate or amenable to upgrading); (3) proximity to transmission lines; (4) resource availability (water as well as wood); and (5) local air-emission dispersion characteristics.

Construction of a 50-MW WFPP would involve significant economic impacts, particularly in the more rural areas. About 500 workers would be directly employed in the two-year construction project, and about 250 jobs would be created indirectly in (1) retail trade (about 100 jobs), (2) wholesale trade (about 25 jobs), (3) medical services (about 25 jobs), and (4) other sectors (about 100 jobs). Unemployment rates would be reduced by about 1.5% in an area such as St. Johnsbury if all these workers were hired locally. Increased construction materials requirements would, of course, be observed with about $5000 in construction lumber, $7500 in concrete slab roofing, and $55,000 in ready-mix concrete going into plant construction each year. Diesel fuel consumed in construction activities would cost about $135,000 per year, 10% of which would be absorbed locally in wholesale trade and delivery services (VanderWerf 1978). These requirements would be significant in the more rural areas.

For the plant sites considered by High (1978), temporary facilities for construction workers are "not likely to be a great problem" because existing facilities within commuting distance appear adequate. Construction noise should also be relatively insignificant because of the plant lot size and distance from population concentrations. Jealousy among citizens could, however, be a real problem as certain incomes are boosted greatly while others remain unaffected (VanderWerf 1978).

The shift from construction to operation of the WFPP will again involve significant dislocations because operating expenditures will be much lower than construction expenditures. Direct employment during operation involves only about 190 jobs (primarily in wood harvesting), and new indirect employment drops to about 100 jobs. Because primary expenditures continue to be for labor, income distribution patterns remain somewhat similar to those of the construction period. Expenditures during operation are, however, more diffuse; for example, some money will go to landowners in the form of stumpage fees for harvested wood. Sharp drops in business in the lodging and construction industries appear likely, particularly in the community in the immediate area of the plant. As expenditures become more diffuse, outlying areas could, in contrast, experience small increases in business with the transition to plant operation. The drop in employment and business in the immediate vicinity of the WFPP at the end of construction probably represents the most serious socioeconomic impact of the plant. Local interests should clearly be informed of construction schedules and probable effects (VanderWerf 1978).

For truck transport of wood chips from the harvest area to WFPP (rail transport is also possible though it is much more capital intensive), diesel fuel consumption would increase by more than a factor of 2 over use during plant construction (VanderWerf 1978). To supply a 50-MW WFPP, 32 trucks carrying 30-ton payloads would have to make 3 ⅓ round trips per day (60-mile average round trip) for a total of 105 truckloads per day. Bridges would have to be replaced in some cases to support such loads, and traffic impacts might require additional climbing or passing lanes. Additional road maintenance requirements are difficult to assess. Noise levels may also present real problems in settled areas, particularly near the power plant site (Adler et al. 1978).

Segerson (1978) indicates that state revenues from income and sales taxes and from truck registrations will exceed state expenditures during both the construction and operation of a WFPP in Vermont. The state would clear, according to her analysis, about $800,000 and $1,250,000 per year during construction and operation, respectively. At current property tax rates, a WFPP would also contribute significantly to local tax pools and could reduce local rates (High 1978).

Institutional relationships with regulatory agencies (utility and environmental), with the existing power grid, and with cogenerated steam customers are not yet completely clear (Hewett and High 1978*a*). Some progress toward resolution of utility issues is being made at a national level (*Inside FERC*, Dec. 3, 1979) and in a few states, including California (*Inside FERC*, Dec. 24, 1979, and Gustafson 1979).

The above discussion focuses on a single forecast biomass technology at a particular scale and in a particular application and local setting. Even if direct combustion turns out to be the preferred forest energy production technology, keeping it within the forest products industry may still be best. Grantham (1977) argues that this industry is already located ideally with respect to the resource; that it is accustomed to dealing with bulky, wet, and sometimes dirty feedstocks; that it has demonstrated an ability to convert wood efficiently to a mix of products and energy; and that it can best absorb available resources. The forest products industry provides its own market for cogenerated steam as well, whereas none of the Vermont towns considered above could currently put "waste" heat to use (Hewett and High 1978*b*). Although potentially similar in terms of wood harvesting activities, type of employment, plant construction requirements, and other impacts, the specific effects of an industrial WFPP are bound to be different from those of a utility electric plant. The industrial plant, for example, might in many cases involve adding to an existing operation at an existing site and might involve very little additional feedstock hauling and other impacts. A preexisting operation of this sort might already be in touch with local opinion and might make better use of local labor and services than could a newly arriving organization. Depending on a variety of factors, including perhaps the size of the concerns involved, the industrial plant might also be more subject to local influence or control than a utility power plant. Specific impacts in both the industrial and the utility cases, of course, would vary with state, forest region, plant size, specific site (community) characteristics, and other parameters.

Other technologies must also be considered. Wood stoves, for example, have already displaced the equivalent of 20, 18, and 27% of residential oil consumption in Maine, New Hampshire, and Vermont, respectively (Bentley 1979), and are receiving emphasis in the Tennessee Valley Authority (TVA) region of the Southeast (Freeman 1979).

Harvesting for space-heating stoves is very different from harvesting for power plant applications. Wood for space-heating applications is often "cut and gathered by individual cutters or small crews" and transported in small (about 2-ton) trucks for local consumption (Groncki et al. 1978). One might imagine that a significant proportion of the wood harvested would be used by the harvester and that wood resources from small landholdings might therefore be available for stoves and not available for power plants. (This harvesting process also tends to be less mechanized and may allow more flexibility in terms of what is cut.) Although the harvesting process is more labor intensive, many of the jobs are of a "nonmarket" nature with no taxable income earned (Groncki et al. 1978). Accident rates will undoubtedly be higher with the less-organized, amateur labor force involved in harvesting for stove use. In addition to accidents with saws and axes, the incidence of heart attacks and other ailments related to unaccustomed exertion is likely to increase (Cambel 1979).

Wood stoves, mobile pyrolysis units (such as the one under development at TVA), and other relatively decentralized technologies do not generally produce the concentrated impacts of a WFPP. Stove or pyrolysis unit production constitutes a relatively conventional industry and does not need to be sited in a rural community. Wood harvesting efforts and associated traffic tend to be more dispersed than is the case with a WFPP, and relatively small or negligible additions to onsite labor are necessary for private home or existing institutional (e.g., college heating plant) applications of the conversion systems. Stoves and pyrolysis units involve a very different sales/distribution/maintenance infrastructure than would a small number of utility-linked WFPPs. Wood storage must also be handled differently, probably at the user's site, and large storage capacities will be required if a mobile unit is to visit a user's site only a few times a year.

Community and regional economic benefits may appear to be smaller for wood stoves than for WFPPs because of nonmarket harvesting for personal use, the lack of tax revenues from the wood stove facility, and the fact that stoves are likely to be imported into the community rather than built locally. On the other hand, although WFPPs in Vermont will not significantly affect the average price of electricity, wood stoves may at least initially provide much cheaper home heating for those willing to provide their own labor. A benefit such as personal satisfaction would not automatically appear on an economic balance sheet. Personal and community lifestyles appear to be more closely affected by home use than by centralized conversion and conventional electricity supply. Not all these effects, assuredly, will be positive—timber thefts, for example, and feelings of resentment between residents and absentee landholders appear more likely with decentralized than with centralized use of wood. Where wood harvesting becomes a seasonal rather than a continuous activity, social costs in the form of increased unemployment compensation may accrue (Hickey 1979). In addition, fire hazards rise with the use of wood stoves (Cambel 1979), and air quality may be more difficult to maintain with many small stoves than with one large electric plant (Sect. 6.5). Current stove designs are also somewhat inconvenient; the physical requirements for putting logs into a stove may be beyond the capabilities of many physically handicapped or elderly people—a serious problem if heavy use of wood reduces availability and raises the price of conventional fuels (Cambel 1979).

The aesthetic impacts alluded to at the end of Sect. 7.1.1 may become a particularly important issue with regard to direct combustion of wood. With over 5 million American families owning wood stoves in the fall of 1979 and more than 1 million stoves sold in the period of a year (Solar Energy Intelligence Report, Oct. 22, 1979), statements such as the following may take on a new meaning:

> Simply stated, the attempt of the population of the eastern United States to live, even at a lower standard, on energy generated from firewood would mean the end within a few generations of the eastern forest (Nash 1979).

> The biomass energy option means the end of the ultimate safeguard of wilderness—the fact that nobody wanted it for anything else (Nash 1979).

In terms of socioeconomic impacts analysis, less attention appears to have been given to the decentralized use of wood than to power plant applications. Further attention needs to be given to harvest management, personal and community impacts, relative local and state revenues, and other aspects of decentralized use. Future competition for feedstocks (and associated feedstock price increases) and the relative merits of forest products cogeneration as compared with utility applications also require further consideration.

7.2.2 Energy Production from Agricultural Biomass—Implications for the Producer

Energy production from agricultural biomass will have direct impacts on farm planting, tillage, and harvest practices. In addition, arrangements for biomass storage and transport will generally be necessary, and biomass conversion facilities and activity (when located in agricultural areas or on farms) will affect jobs and farm incomes. Some potential impacts at the farm are outlined below. Further details of farm impacts are also contained in the methane and ethanol discussions that follow.

Cropping practices—that is, summer fallow, crop rotation, double cropping, etc.—are often carefully coordinated to maximize profit within equipment, labor, fertilizer cost, and other constraints on U.S. farms (Tinberg and McElroy 1978 and Abeles and King 1978). Biomass production will add to the complexity of this process and will result in secondary production impacts as well as direct crop substitutions. Where a farmer currently plants equal amounts of corn and soybeans, for example, he might switch to an emphasis on corn, given a residue biomass market, because of the relatively low biomass yields of soybeans (Benson et al. 1978). On a 320-acre Iowa farm, a minimum of 200 acres of corn is required by the harvest and processing economics of stover for confinement livestock feed, so that tentative or experimental crop shifts will not be practical in this case. On the same farm, variable loading problems with a methane digester tend to favor the year-round farrowing of pigs over beef operations; the digester may stand empty between buying and selling (Iowa State University 1977). Stoichiometric situations such as these could lead to problems if too many independent farmers make the same choice. Shifts in crop rotations from corn-corn-soybeans to continuous corn or from corn-oats-meadow-meadow to corn-corn-oats-meadow will also have a direct effect on fertilizer energy (Iowa State University 1977), labor, equipment, and other requirements. Although farmers are generally accustomed to a multiplicity of objectives and constraints, sufficient incentives will have to be provided to make these additional complexities worthwhile.

Agricultural biomass harvesting equipment is not very well developed and is likely to involve a significant capital investment (Alich et al. 1977). Technical and economic risks and low biomass revenues on a per-acre basis are likely to make the purchase of harvesting equipment difficult or place it low on the farmer's priority list and may favor contracted harvesting (Tinberg and McElroy 1978). Arrangements of this sort, however, are likely to be restricted by prior experience with the field damage and harvest delays often associated with such services (Miles 1977). Interestingly, residue ownership may present a problem (Alich et al. 1977) under certain existing contractual agreements (e.g., with tenant farmers).

The economics, as well as the energy requirements, of residue utilization will be strongly affected by the extent to which residues can be harvested along with primary products and by the seasonal or perishability characteristics of the crop (Miles 1977). Climatic conditions will probably be important in this regard. Illinois, for example, experiences 26 days of precipitation in the October–December quarter (stover harvest season), allowing a maximum equipment utilization of only about 660 h/year (Tinberg and McElroy 1978). Local labor conditions and the farmer's other seasonal commitments may therefore affect his interest in biomass production for energy. Paid versus unpaid or hourly versus task pay distinctions may be important. Exaggeration of the seasonal characteristics or migratory work-force impacts in rural communities may present both practical problems (e.g., requirements for housing and other facilities that are used for only a short time each year) and questions of social palatability.

The transportation requirements will be strongly affected by the scale and location of conversion facilities and again by the seasonal and perishability characteristics of the biomass crop. Seasonal crops will probably be stored mainly at the farm site in order to diffuse the rush for labor and transport equipment at harvest time and to reduce storage requirements at the conversion plant (Miles 1977). Land requirements for storage periods of six months and more are not unusual (Tinberg and McElroy 1978), and whenever possible, the harvesting process should "roadside" materials in a form that is easy to move to the conversion facility (Miles 1977). Land requirements for storage at the farm do not appear to present a problem, although difficulties are more easily imagined in cases where sheltered storage (some sort of structure) is necessary for protection from the weather. Erratic demands and specialized equipment requirements may result in the necessity for a conversion company to operate its own biomass transport equipment rather than to pay for either contract or common carrier services (Miles 1977). In some cases, however, farmers will probably be required, or will volunteer, to deliver their own biomass harvests.

Storage of both seasonal and continuously harvested biomass resources will have implications for the location and execution of intermediate processing activities (drying, grinding, etc.) where applicable, and consideration of dust explosions and spontaneous combustion will be necessary in some cases (Miles 1977). [See Barbera (1979) for a discussion of the interaction of wood chip storage, drying, and transportation processes.] Storage losses [e.g., ground contact losses average about 6% for open storage of roll-baled forage sorghum (Tinberg and McElroy 1978)] and local climatic effects [e.g., round bales do not weather well in damp climates (Miles 1977)] may also have important implications. Although generally avoiding difficulties of this sort, continuously harvested crops may present other problems; for example, kelp, hyacinths, and algae (to stretch the "agricultural" heading a bit) are all perishable and require management of harvesting and transportation so that they can be processed within one to three days (Miles 1977).

Details of the demands placed on farmers by agricultural biomass will depend on the specifics of the crop and the conversion technology to be employed. It is clear, however, that energy production from agricultural biomass will add significantly to the farmer's burdens in many cases. Already saddled with a multiplicity of roles, responsibilities, and uncertainties, farmers may be acutely aware of additional expertise, equipment, and time that are necessary both in handling and in managing biomass systems and may wish to avoid new constraints and complications in optimizing overall farm operations. These impacts will probably play a major and, in some cases, a dominant role in the choice between adoption and nonadoption of new biomass energy options.

Although they have received little attention to date, several other aspects of agricultural biomass development need to be examined carefully. Community, state, and regional impacts, including employment, skill requirements, and available labor, in particular, require investigation. Other questions include the following: What effects on community life can be expected from local conversion facilities and intensified or extended seasonal activity? Are cultural disparities between rural and urban areas likely to be accentuated by rural use of biomass? What wage levels are to be expected and how will they influence the mix of local and imported as well as permanent and temporary workers? Will biomass resources be used locally or converted to fuels for export from producing areas, and what local and regional economic implications would be associated with each of these options?

7.2.3 Methane—Combined Production and Conversion

The economic attractiveness of methane production from confinement feedlots and dairy operations is greatly enhanced by an existing need for manure handling and disposal systems. Where handling systems compatible with methane digestion are adopted [hydraulic systems are well suited because water is added to manure for digestion (Iowa State University 1977)], anaerobic methane production can contribute greatly to solving both the odor and the storage and disposal problems that face farms operators (Iowa State University 1977, Varani et al. 1977, and Braithwaite and Capener 1978). In most cases, methane production would also provide self-sufficiency for the farm in terms of direct energy inputs [something that many farmers apparently value highly (Braithwaite and Capener 1978)] and would normally result in excess gas as a marketable product. Finally, recycling of digester sludge promotes efficient reclamation of nutrients, including nitrogen (Braithwaite and Capener 1978, Iowa State University 1977), and would reduce farm fertilizer costs. These benefits are, however, subject to certain constraints and may involve undesirable side effects.

Smaller farms in particular must consider the management, labor, technical expertise, and investment demands of digester operation as well as system reliability (Braithwaite and Capener 1978). Indications are that farm digesters will be much less temperamental than pollution control digesters (Iowa State University 1977), although antibiotics used in animal husbandry may lead to some of the complications of pollution-control toxicity effects. Confinement operations are already characterized by intensive management requirements (Iowa State University 1977), and labor needs beyond those associated with the manure handling system can be relatively small (Skrinde 1978). On the other hand, practical difficulties involved in returning digester sludge to the land militate in favor of refeeding the sludge—a process that can claim twice as many man-hours as grain feeding (Iowa State University 1977). With regard to capital investment, two points may be raised: (1) although farmers are generally accustomed to sizable investments and will in any case require manure handling facilities for confinement feeding, methane production systems may be viewed as a low-priority item (Braithwaite and Capener 1978); and (2) financial institutions could be reluctant to lend money for what might be considered an unproven methane production technology (Thornton 1978). Government action in terms of (1) environmental regulation of manure disposal and (2) technology development and demonstration could be very useful in this regard.

Methane generation appears to be more attractive in large than in small confinement operations (Iowa State University 1977, Braithwaite and Capener 1978). Unfortunately, major concentrations of large feedlots are often located in regions characterized by water shortages. Water consumed in liquefying manure for digestion may therefore present a significant constraint (Braithwaite and Capener 1978), particularly in western feedlots. Methane digestion could also enhance the competitive position of large confinement operations relative to smaller "ruminant" farms, posing a further threat to the already disappearing smaller operations (Braithwaite and Capener 1978).

A major area of uncertainty, to some degree unique to combined production and conversion enterprises, exists with regard to the interface between methane-producing farmers or ranchers and gas or electric utilities. No more than 40% of the methane produced is required for heating the digester, and on small farms (those with 100 to 200 cows) the gas from about 15 cows is sufficient to provide space heating, cooking, and hot water for the farmhouse. To be used, excess gas must be transported to another user. Even if it is used at the farm site to produce electricity, an involved process in itself, a perfect match between electric production and demand is highly unlikely and an excess is again likely (Skrinde 1978). Furthermore, gas and/or electric output will vary with digester loading, with season [digester and other heating loads (Varani et al. 1977)], and other factors. Even where existing pipeline or electric utility connections are adequate, the management, scale, quality control, and distribution characteristics of utilities and biomass methane systems are not well matched. Compressed gas storage systems may be appropriate in some cases and would allow for bulk collection of methane by truck in a manner similar to current milk collection systems; associated pressurization equipment would, however, be expensive and extensive safety precautions would be necessary. Electrical distribution systems, like gas systems, do not easily accommodate the addition of excess power even from relatively large power producers. Conversion of farm equipment, primarily tractors, to burn digester gas may be worth considering on small farms (Skrinde 1978) to avoid utility interface problems but would not absorb excess production at large feedlots.

Farmers may to some degree view the methane digester as yet another new and complex hassle. At present, the major trade-off for the farmer appears to be between the digester's primary value as a waste handling mechanism (Braithwaite and Capener 1978) and the major difficulties involved in making use of the gas that is generated (Skrinde 1978). Therefore, careful attention needs to be given to the marketing of excess gas production. The specific social and economic impacts of large-scale methane production on communities and at the state and regional levels have, again, received too little attention. Analysis to the level of detail of the Dartmouth wood-fired power plant study would seem well advised, particularly for locations characterized by major concentrations of large feedlots such as those in the West.

7.2.4 Ethanol—Market Development and Large-Scale Subsidies

It has been said that ethanol production from biomass and its descendent, the gasohol business, "is flourishing like no other segment of the solar community. . . ." About 700 gasohol outlets began operation in the Midwest between summer 1978 and April 1979, and outlets may soon begin to appear in states like Virginia, Pennsylvania, New Jersey, and New York (Solar Energy Intelligence Report, Apr. 30, 1979). Major subsidies and rapid market developments have made ethanol one of the first biomass energy technologies to have an impact on the social and economic life of the nation.

As the gasohol market expands, related industries will feel important impacts. One publication (U.S. Department of Agriculture 1978) suggests that a major ethanol program [i.e., 3.78 million m^3 (10 billion gal) of ethanol per year] could involve sharp reductions in soybean production caused by land transfers into grain production and that the use of distiller's dried grains "would be enough to depress soybean oil meal prices to the extent that the domestic soybean crushing industry would be largely supplanted. . . ." Use of inexpensive distiller's grains in cattle feed would also retard livestock weight gain, would potentially reduce aggregate meat production, and would raise consumer prices and improve farmer profit margins slightly. Food grain prices would also be expected to rise in the program outlined by the U.S. Department of Agriculture.

Ethanol for industrial applications is currently produced as a by-product of petroleum refining. Current producers will probably have little difficulty in preventing competition from new competitors; in fact, current producers are already faced with substantial excess capacity (U.S. Department of Agriculture 1978, Anderson 1978). The distribution of biomass ethanol among beverage and fuel markets could, however, have implications for the beverage alcohol industry, human health, tax revenues, and beverage law enforcement (Oak Ridge National Laboratory 1979). As a sample of possible interactions, Archer Daniels Midland has recently announced that it will convert its 68,000-m^3 (18 million gal) per year plant at Decatur, Illinois, from beverage to anhydrous ethanol production for fuel (Solar Energy Intelligence Report, April 23, 1979).

Finally, and perhaps most obviously, impacts on the gasoline industry are to be expected in terms of both production (refinery product mix, total output, etc.) and distribution. In Brazil, ethanol and gasoline have been found to be adequately mixed in a tank truck when they are added separately if the truck travels at least 64 km (40 miles) to its delivery point (Solar Energy Intelligence Report, Apr. 23, 1979). Separate storage tanks will be used, however, at various points in the distribution network, and additional pumps, mixing systems, and inventory complications can also be expected (Solar Energy Intelligence Report, Apr. 30, 1979, Yang and Trinidade 1978, and U.S. Department of Agriculture 1978). Where alcohol is to be used as a transportation fuel, oil companies would generally prefer methanol production from coal over biomass ethanol (Anderson 1978), partly because of their major ownership in coal resources.

The community and personal impacts of locating ethanol conversion facilities in rural areas would probably be significant. Income effects and labor availability in terms of both man-hours and mix of skills may be at issue, particularly for the smaller-scale, local conversion facilities (Abeles and King 1978). One of the avowed purposes of the Brazilian ethanol program is the creation of jobs at the farm level and the consequent possible improvement in income distribution and more balanced development across the country. Success in this regard has been somewhat limited, and it appears that rural income benefits from conversion facilities may require additional stimulation of local initiative, at least in that country (Yang and Trinidade 1978). Long-term income and farm practice effects caused by unexpected erosion or slowed livestock cycles can also be imagined. Although large conversion facilities might have smaller impacts in more industrialized locations, the scale of conversion facilities will have a major impact on transportation and energy requirements, both in terms of transportation energy consumption and in terms of the need to dry perishable distillation residues for cattle feed (Peters 1979 and Abeles and King 1978). Market variations resulting from weather, farmers' planting decisions, cattle feed price changes, or locally "glutted" end-product markets are possible (Abeles and King 1978, U.S. Department of Agriculture 1978) and militate in favor of farmers' cooperatives and diversification of feedstocks and end products (Abeles and King 1978 and Black 1977). Regional shifts in livestock production (to take advantage of low-cost feed produced at distilleries) could also occur (U.S. Department of Agriculture 1978).

Federal and state incentives for ethanol production from biomass are substantial, amounting in Nebraska, for example, to more than 9¢/gal of gasohol or 90¢/gal of ethanol in gasoline tax exemptions (Anderson 1978 and Maullin et al. 1979) and including grants, loans, load guarantees, and other forms of assistance. New incentives, however, continue to be adopted: recent regulatory changes allow refiners to spread the cost of alcohol over all their gasoline and to increase prices by 2¢ for each additional gallon of unleaded gas (used in making gasohol) they produce (Solar Energy Intelligence Report, November 26, 1979, p. 474). Although these rapidly developing incentives do stimulate production and distribution, even among major oil companies (Solar Energy Intelligence Report, November 26, 1979, p. 476), they raise some serious questions.

1. How much subsidized development is necessary to determine ultimate costs and energy flows?
2. What continuing level of subsidy (if any is necessary) is appropriate to the perceived benefits of biomass ethanol production, and what might the impacts of future subsidy reduction be?
3. Under current subsidies for fuel applications, will users of alcohol for conventional products be forced to pay more in competition for available supplies (Samuels 1979)?
4. If more and more states begin to offer tax subsidies for gasohol to increase their fuel supplies, could subsidy wars develop between states over limited alcohol supplies (Samuels 1979)?

Incentives have become so extensive, and the development of some alternatives has been so rapid, in fact, that the Department of Energy (DOE) has joined the Department of Agriculture in openly opposing further legislation promoting specifically wood and forestry or agricultural waste energy production. In testimony to this effect before two congressional committees, Bennett Miller of DOE suggested a "need to exercise caution" in developing these resources in order to ensure that they are harnessed "without harm to ... ecological, recreational, and traditional nonenergy needs..." (Solar Energy Intelligence Report, Nov. 26, 1979, p. 473).

At current levels, total subsidies could well rise into the billions of dollars per year for a major gasohol commitment (see, for example, U.S. Department of Agriculture 1978). The effects of incentives of this magnitude should be monitored carefully.

7.2.5 An Added Note

As a final note, agricultural biomass energy contributions clearly depend heavily on farmer attitudes. "... Proposals to use alcohol originat(ing) from the desire of farm groups to develop an added market for surplus grains" (U.S. Department of Agriculture 1978) appear to have played an important role in moving this biomass technology toward implementation over significant technical and economic concerns. Attitudes such as those displayed by farmers in a small Pratt, Kansas, survey, on the other hand, could significantly restrict the pace and extent of agricultural biomass energy development: 60% of the 154 responding farmers indicated that they would not allow removal of any crop residue from their lands, primarily because of erosion control concerns. The 40% who were willing to allow residue removal were interested in only partial removal from only portions of their land holdings (Koelsch et al. 1977). Although these results reflect traditional education and long-standing practice, farmers do generally control residue supplies and crop plantings. Because they are certain to be closely involved with harvest, transport, and even conversion of agricultural biomass to energy, their eventual cooperation is a virtual necessity to successful implementation of agriculturally based biomass programs. As already indicated, similar comments may apply to the harvesting of wood from privately owned lots and to other biomass energy sources.

7.3 Long-Term Technologies

Almost no research appears to have been done on the possible socioeconomic impacts of long-term biomass options such as monocultural coppicing tree farms or open-ocean kelp farms. As currently conceived, these options are characterized by more intensive management for higher yields and, in some cases, are much larger in scale. As an initial approximation, one might expect their impacts to be correspondingly intensified. Major new impacts will also need to be considered, however. Kelp farming, for example, could involve ocean buoy, nylon rope, kelp harvesting ship, and other component requirements that would essentially absorb total current production of such components (Wilcox 1976). Farms of several hundred square miles could involve critical issues of coastal zone development (cf. offshore oil impacts), potential storm damage, and effects on current maritime commerce (Sect. 5.5).

The environmental impacts of advanced, intensively managed biomass farming concepts in a sense become social impacts in themselves. The lack of diversity in the wildlife population of a monocultural silviculture farm (Sect. 5.1.1.4) and the possibility of CO_2 releases from deep ocean waters associated with open-ocean kelp farming (Sect. 5.5), for example, could have direct and powerful social implications. The unprecedented scale and pervasiveness of resource management implied in some of the advanced concepts may require some social or psychological adjustment in and of itself.

With public consideration only beginning, indications already show that massive, long-term biomass farming concepts may be publicly unacceptable under current circumstances. In a recent DOE-funded community technology assessment, citizen volunteers in a three-county region of New York state set as one of four limits on their energy future what they called "ecological balance."

> It is not enough to replant with one variety of tree after clear-cutting a forest composed of oak, beech, maple, and other deciduous trees, nor to replant row upon row of white pine where there once stood a forest of many species of coniferous evergreens.

In eliminating such options from consideration for their region, these citizens suggest that the reader "...contemplate what the impact on American forests would have been if prior to 1900 all clear-cut forests had been reseeded with American elm or American chestnut," both of which, we are reminded, "have been nearly extinguished by non-native disease" (Banfield et al. 1979).

Public perceptions will vary with location and with time. It is clear, however, that very substantial public commitments would be required for the implementation of large-scale biomass farming concepts. Assessment of the social and economic impacts of such concepts is critically needed and should not be further postponed for options under active consideration. Frank and open consideration should be given to the current relatively unspoiled nature of potential biomass farming sites and to the demographic and economic dislocations that might be entailed in opening up a large-scale farm. Other possible uses for the land, water, and other resources absorbed by such a farm also need to be considered along with their respective social costs and benefits. Whether or not the sole object of a farming concept is the production of energy, other sources of energy need to be carefully considered (including, for example, other renewable solar technologies). Thought should also be given to the reversibility of near-term decisions and to the implications of possible long-term trends (desertification or other ecological disasters) that are only dimly perceived at present.

To the extent that current popular support for biomass energy systems hinges on a reliance on undisturbed natural cycles, intensive management concepts may fall short. Careful monitoring of the development of more modest near-term applications could provide useful clues as to the social palatability of long-term alternatives and indicate the level of support that might eventually be obtained for such alternatives. Direct public involvement in assessment of long-term biomass options could also be very useful.

7.4 Ethics and Aesthetics

The aesthetic attraction of life in an untouched wilderness has long been a part of our culture, surfacing frequently in classic American literary works (*Walden, Moby Dick, Huckleberry Finn,* etc.) and tracing its literary roots back through Shakespeare to Theocritus and Virgil. Classically, the theme of a return to an unspoiled natural setting often subsumes a retreat from the complexities of society and (more recently) technology (Marx 1973). In modern terms, what has been called the "cult of compulsive manipulation" has served as a springboard for the "cult of the simple rustic life" (Santmire 1973) with its definite "environmentalist" and "back to nature" components.

The appeal of the land, of nature, is not simply aesthetic in the sense that one may, for example, prefer blue over green in selecting a tie or hat. Aesthetics with regard to land and land use appears to have clear spiritual and *ethical* components. "In Israel thousands of years ago, Isaiah cried out to his people: 'Woe unto them that join house to house, that lay field to field, till there be no place, that they may be placed alone in the midst of the earth!' " (Sierra Club 1980 and Isaiah ch. 5, verse 8). As early as the 1940s, Aldo Leopold, in a much quoted passage, foreshadowed the evolution of an environmental ethic which was at least partly realized in the late 1960s and early 1970s:

> The first ethics dealt with the relation between individuals; the Mosaic Dialogue is an example. Later accretion dealt with the relation between the individual and society. The Golden Rule tries to integrate the individual to society; democracy to integrate social organization to the individual.
>
> There is as yet no ethic dealing with man's relation to land and to the animals and plants which grow upon it. Land, like Odysseus' slavegirls, is still property. The land-relation is still strictly economic, entailing privileges but not obligations.
>
> The extension of ethics to this third element in human environment is, if I read the evidence correctly, an evolutionary possibility and an ecological necessity (Leopold 1949).

Shortly before the oil embargo of 1973, C. D. Stone proposed a legal expression of such an environmental ethic in which natural objects would be vested with rights similar to those of citizens and corporations: that is, the right to (1) institute legal actions, (2) seek legal relief based on personal harm (harm to the natural object itself rather than a third party), and (3) relief from damages which would benefit the river, forest, etc., directly. While these rights still have not been explicitly recognized (in many cases, a "third party" must be economically damaged before the pollution of a river, for example, becomes a legal issue), similar ends have to some degree been achieved through other approaches (Stone 1972).

Energy and economic problems appear in recent years to have displaced environmental issues as the primary concerns of our society. Many social observers, however, view the energy situation as simply another manifestation of the problems that underlie environmental and other preexisting issues. Even before the energy crisis, some referred to doubts about economic growth, resource limitations, and global economic imbalances in the same breath with the dangers of irreversible ecological damage (Ward 1973). In a book entitled *Bioethics/Bridge to the Future*, Van Rensselaer Potter suggested in 1971 a new concept of progress, that is, progress in the knowledge of how to use knowledge, which may be an even more appropriate alternative today to the conventional material concept than it was before the 1973 oil embargo (Potter 1971). More recently, Harman combined energy/resource and growth concerns with environmental, work satisfaction, world distribution, and other dilemmas to suggest a single coherent pattern and a possible transformation of society from that pattern (the "industrial era paradigm") to a renewed expression of more basic values (a "trans-industrial" society) (Harman 1979).

Others (Harman 1979; Schwartz 1979; Elgin and Mitchell 1977) would argue that we are approaching a historic turning point in our history—perhaps *the* point of choice characterized by Leopold as "an evolutionary possibility and an ecological necessity" (Leopold 1949). Our development of biomass energy systems is perhaps just one of many tests of our "knowledge of how to use knowledge" as Potter (1971) would say, or, to use Leopold's term, of our "scholarship."

> Ability to see the cultural value of wilderness boils down, in the last analysis, to a question of intellectual humility. The shallow-minded modern who has lost his rootage in the land assumes that he has already discovered what is important; it is such who prate of empires, political or economic, that will last a thousand years. It is only the scholar who appreciates that all history consists of successive excursions from a single starting point to which man returns again and again to organize yet another search for a durable scale of values. It is only the scholar who understands why the raw wilderness gives definition and meaning to the human enterprise (Leopold 1949).

Though difficult to deal with in a technical and environmental context, aesthetic and ethical issues are of transcendent human concern. In the last analysis they can neither be dismissed nor ignored.

7.5 Conclusions

Extensive development of biomass as an energy source is expected to precipitate severe land-use conflicts and an extensive restructuring of the nation's rural life. Major transportation impacts and growth in rural employment may lead to a variety of secondary effects (increasing suburbanization, growing urban/rural cultural homogeneity or disparity, etc.), depending on how the former impacts are handled. Other potentially critical issues include the major expansion of farmers' roles to include responsibility for complex energy production and conversion systems and the extraordinary institutional interface problems associated with decentralized production of gas and electricity for sale to utility networks.

As we begin to analyze the initial impacts of a few technologies in a few locations, we may begin to ask more intelligent questions about other technologies and about the future. What new impacts can be expected in other locations or with other technologies? How will the effects of many installations of various designs over a long period of time compare with the initial effects of single installations? In addition, programs can begin to be developed to consider the relative merits (in terms of social and energy-efficiency benefits as well as pure economics) of centralized and decentralized applications. Monitoring and projection efforts can be initiated to prevent overexpansion and wasteful practices and to begin sorting out the technologies that will be viable without subsidy on a long-term basis from those that, as a result of market variations or other factors, are likely to remain net dollar or energy sinks. Lessons may also be drawn from near-term applications (e.g., regarding the attractiveness of multiple benefits such as food and energy or energy and manure disposal) for use in the design of new, long-term technologies. National, regional, and local economic and political patterns, international trade, food and conventional energy production, and a host of other areas will need to be taken into account in addition to those mentioned above.

In considering the full range of biomass options (both near and long term), energy and environmental issues raise increasingly prominent and very fundamental philosophic, aesthetic, and practical questions: To what extent is nature to be exploited for human purposes? What is the purpose of "domestication?" (i.e., What are our needs?) How are conflicting desires (e.g., wilderness preservation versus forest energy development) or the needs of different groups (e.g., energy versus food) to be reconciled? How are potential future costs and benefits to be weighed against current costs and benefits, and how are "noneconomic" factors to be internalized in this process? In our judgment, growing public awareness of the social and economic implications of biomass energy production and of the energy returns for a given commitment will restrict development to the more moderately conceived applications. Within that limitation, however, lies a great range of possible response to the questions outlined above. Further analysis and discussion of potential specific socioeconomic impacts would greatly facilitate the difficult choice-making process that we collectively face.

References

Abeles, T. P., and J. R. King, 1978. *Parameters for Legislative Consideration of Bioconversion Technologies*, NSF/RA-780196, National Science Foundation, Washington, D.C.

Adler, T. J., M. Blakey, and T. Meyer, 1978. *The Direct and Indirect Costs of Transporting Wood Chips to Supply a Wood-Fired Power Plant*, report prepared under U.S. Energy Research and Development Administration Contract EG-77-C-02-4487, Dartmouth College, Hanover, N.H.

Alich, J. A., et al., 1977. *An Evaluation of the Use of Agricultural Residues as an Energy Feedstock—A Ten-Site Survey*, vol. 1, *Summary and General Information*, report prepared under ERDA Contract E(04-3)115, Letter Agreement No. 109, Stanford Research Institute, Menlo Park, Calif.

Alich, J. A., and R. E. Inman, 1976. "Energy from Agriculture—The Most Economic Method of Large-Scale Energy Conversion," *Energy* **1:** 53-61.

Anderson, E. V., 1978. "Gasohol: Energy Mountain or Molehill?", *Chem. Eng. News* **56:** 8-13.

Banfield, B., et al., 1979. "Plan for Local Energy Conservation and Development," preliminary version developed and written by volunteers from Chemung, Steuben, and Schuyler counties, N.Y.; published by the Southern Tier Central Regional Planning and Development Board, Corning, New York, under funding from the Department of Energy, Washington, D.C.

Barbera, N. J., 1979. "Optimum Methods of Handling Woodchips for Energy Plantations," in *Proceedings of the 14th Intersociety Energy Conversion Engineering Conference, Aug. 5-10, Boston, Mass., American Chemical Society, Washington, D.C.*

Bennett, C. F., 1979. "Some Social and Economic Consequences and Constraints to the Use of Forests for Energy and Organics in Canada, the United States of America, and Los Estados Unidos De Mexico," in *Biological and Sociological Basis for a Rational Use of Forest Resources for Energy and Organics / An International Workshop for the Man and the Biosphere Program, Final Agenda*, Michigan State University, East Lansing.

Benson, W., et al., 1978. *Systems Study of Fuels from Grains and Grasses, Phase 1, Final Report*, ALO/3729-1, U.S. Department of Energy, Washington, D.C.

Bentley, G. K., 1979. "New England Rural Energy Sources—Wood, Small-Scale Hydro, Direct Solar and Wind—An Assessment of Near-Term Potential," New England Congressional Caucus, Washington, D.C.

Berger, G. J., 1977. *The Legal and Regulatory Framework for the Interface of a Wood-Fired Power Plant and the Electric Power Grid in Vermont,* Rep. DSD 95, Dartmouth College, Hanover, N.H.

Black, W. E., 1977. "Alternative Organizational and Marketing Arrangements for Marketing Biomass," in *Biomass—A Cash Crop for the Future?* Rep. CONF-770368. U.S. Department of Energy, Washington, D.C.

Blake, D., and D. Salo, 1977. *Sytems Descriptions and Engineering Costs for Solar-Related Technologies,* vol. IX, *Biomass Fuels Production and Conversion Systems,* Rep. MTR-7485, U.S. Energy Research and Development Administration, Washington, D.C.

Braithwaite, D. C., and H. R. Capener, 1978. "Social Considerations of Methane Production on Dairy Farms," Chap. 9 in *Anaerobic Fermentation of Agricultural Residue: Potential for Improvement and Implementation*, Rep. HCP/T2981-07 UC-61, U.S. Department of Energy, Washington, D.C.

Cambel, A. B., 1979. *Social Assessment of On-Site Solar Energy Technologies*, Rep. HCP/R4040-02, U.S. Department of Energy, Washington, D.C.

Chambers, R. S., et al., 1979. "Gasohol: Does It or Doesn't It Produce Positive Net Energy?" *Science* **206:** 789–95.

DiNovo, S. T., et al., 1978. *Preliminary Environmental Assessment of Biomass Conversion to Synthetic Fuels*, Rep. EPA-600/7-78-204, U.S. Environmental Protection Agency, Cincinnati, Ohio.

Edminster, T. W., 1976. "Bioconversion By-products: Feeds, Fertilizers, and Feedstocks," in *Capturing the Sun Through Bioconversion*, Bio-Energy Council, Washington, D.C.

Elgin, D., and Mitchell, A., 1977. "Voluntary Simplicity (3)," in *The Coevolution Quarterly*, Summer/1977, Sausalito, Calif.

Energy Insider, Apr. 2, 1979. "BPA Studies Wood as Plant Fuel," *Department of Energy Newsletter* **2**(7), Washington, D.C.

Freeman, S. D., 1979. Keynote address: "On the Rise 2" session, 1979 Congress, International Solar Energy Society, Atlanta, Ga., May 29, 1979.

Grantham, J. B., 1977. "Anticipated Competition for Available Wood Fuels in the United States," in *Fuels and Energy from Renewable Resources,* D. A. Tillman et al., eds., Academic Press, New York.

Groncki, P. J., et al., 1978. *Assesing the Employment Implications of Alternative Energy Supply, Conversion and End Use Technological Configurations: The Case of Firewood Versus Fuel Oil in New England in 1985*, Rep. BNL 24048, U.S. Department of Energy, Washington, D.C.

Gustafson, L., Assistant to Claire Dedrick, California Public Utilities Commission, personal communication, Oct. 25, 1979. See also recent staff exhibits of the California P.U.C., including Exhibit No. 41, "Recommended Commission Policies and Price Rules for Utility Purchases of Cogenerated, Auxiliary and Small Production Facility Power" (7/20/79), and other exhibits issued in response to P.U.C. Order Instituting Investigation Number 26.

Harman, W. W., 1979. *An Incomplete Guide to the Future*, W. W. Norton & Company, New York.

Herendeen, R. A., 1978. "Energy Analysis of Two Technologies: Gasohol and Solar Satellite Power Station," in *Symposium Papers, Energy Modeling and Net Energy Analysis*, Institute of Gas Technology, Chicago, Ill.

Hewett, C. E., 1978, *The Availability of Wood for a 50-MW Wood-Fired Power Plant in Northern Vermont*, Rep. DSD 114, prepared for U.S. Department of Commerce under grant 01-6-01659 by Dartmouth College, Hanover, N.H.

Hewett, C. E., and C. J. High, 1978*a*. *Construction and Operation of Small, Dispersed, Wood-Fired Power Plants*, testimony prepared as Rep. DSD 117A for Energy Resources Conservation and Development Commission of the State of California, Dartmouth College, Hanover, N.H.

______, 1978*b*. *The Feasibility of Operating a 50-MW Wood-Fired Power Plant in Northern Vermont: Wood Availability and Siting Potential—Executive Summary*, Rep. DSD 116, prepared for U.S. Department of Commerce under grant 01-6-01659 by Dartmouth College, Hanover, N.H.

Hickey, H. R., Chief, Applied Research Staff, Tennessee Valley Authority, personal communication with H. Braunstein (ORNL), Sept. 10, 1979.

High, C., 1978. *Evaluation of Locations for a 50-MW Wood-Fired Power Plant in Northern Vermont*, Rep. DSD 115, prepared for U.S. Department of Commerce under grant 01-6-01659 by Dartmouth College, Hanover, N.H.

Hudson, W., 1977. "Commercial Use of Corn Cob Residue," in *Biomass—A Cash Crop for the Future?* CONF-770368, U.S. Department of Energy, Washington, D.C.

Hunter, Y. L., et al., 1976. "The Long Term Implications and Constraints of Alternate Energy Policies," *Energy* **1:** 375–406.

Inman, R. E., 1977. *Silvicultural Biomass Farms—Summary*, MTR-7347 (vol. 1), U.S. Energy Research and Development Administration, Washington, D.C.

Inside FERC, Dec. 3, 1979. McGraw-Hill, Inc., New York.

Inside FERC, Dec. 24, 1979. McGraw-Hill, Inc., New York.

Iowa State University, 1977. *Energy Relationships for a 320-Acre Iowa Farm—An Attempt to Increase Production of Grain and Meat While Rendering the Farm Energy Self-Sufficient by Generating Methane Through Anaerobic Fermentation of Residues*, PB-275 274, U.S. Department of Commerce, Washington, D.C.

Koelsch, R. K., et al., 1977. "Use of Crop Residues to Support a Municipal Electrical Utility," in *Food, Fertilizer and Agricultural Residues, Proc. 1977 Cornell Agricultural Waste Management Conference*, R. C. Loehr, ed., Ann Arbor Science Publishers, Inc., Ann Arbor, Mich.

Leopold, A., 1949. *A Sand County Almanac and Sketches Here and There*, Oxford University Press, London.

Lipinsky, E. S., 1978. "Fuels from Biomass: Integration with Food and Materials Systems," *Science* **199:** 644–51.

Marx, L., 1973. "Pastoral Ideals and City Troubles," in *Western Man and Environmental Ethics*, ed. by I. G. Barbour, Addison-Wesley, Reading, Mass.

Maullin, R. L., et al., 1979. "Energy Choices for California...Looking Ahead," introduction to the *1979 Biennial Report of the California Energy Commission*, State of California, Sacramento.

McAllister, P., 1976. "International Aspects of Biomass Conversion," in *Capturing the Sun Through Bioconversion*, Bio-Energy Council, Washington, D.C.

McElroy, A. D., 1977. "Utilization of Land with Limited Capabilities," in *Biomass—A Cash Crop for the Future?* CONF-770368, U.S. Department of Energy, Washington, D.C.

Miles, T. R., 1977. "Logistics of Energy Resources and Residues," in *Fuels and Energy from Renewable Resources*, D. A. Tillman et al., eds., Academic Press, New York.

Nash, Roderick, 1979. "Problems in Paradise," in *Environment* (July/August).

Oak Ridge National Laboratory, 1979. *Environmental Assessment of an Alcohol Fuels Program*, ORNL NEP-II Impact Assessment Project for the U.S. Department of Energy, Oak Ridge, Tenn.

Olson, K. W., 1977. "Social Returns from Liquidating Excess Inventories of Public Timber," *Land Econ.* **53(3):** 342–58.

Peters, J. F., April 18, 1979. University of Alabama, Huntsville, 1979, personal communication.

Pimentel, D., et al., 1975. "Energy and Land Constraints in Food Protein Production," *Science* **190:** 754-61.

______, 1976. "Land Degradation: Effects on Food and Energy Resources," *Science* **194:** 149–55.

Potter, V. R., 1971. *Bioethics/Bridge to the Future*, Prentice-Hall, Inc., Englewood Cliffs, N.J.

Reuyl, S. E., et al., 1977. *Solar Energy in America's Future—A Preliminary Assessment* (2d ed.), Rep. DSE-115/2.

Samuels, G., 1979. Personal communication to J. Tatum, July 2, 1979.

Santmire, H. P., 1973. "Historical Dimensions of the American Crisis," in *Western Man and Environmental Ethics*, ed. by I. G. Barbour, Addison-Wesley, Reading, Mass.

Schwartz, P., 1979. "Luncheon Address," in *Proceedings of the Third Stationary Source Combustion Symposium*, vol. V, Addendum, EPA-600/7-79-050e.

Segerson, K., 1978. *Vermont State Revenues and Expenditures Generated by a 50-MW Wood-Fired Power Plant*, Rep. DSD 109, Dartmouth College.
Sierra Club, 1980. "Wilderness 1980/Sierra Club Engagement Calendar," Charles Scribner's Sons, New York.
Skrinde, R. T., 1978. *Engineering Feasibility of Fuels from Biomass Utilizing Dairy Farm Residue*, prepared under contract ET-78-X-01-2404-0 for the U.S. Department of Energy, Washington, D.C.
Solar Energy Intelligence Report, Apr. 30, 1979. Business Publishers, Inc., Silver Spring, Md., A. L. Frank, ed.
Solar Energy Intelligence Report, Mar. 12, 1979. Business Publishers, Inc., Silver Spring, Md. A. L. Frank, ed.
Solar Energy Intelligence Report, Apr. 23, 1979. Business Publishers, Inc., Silver Spring, Md., A. L. Frank, ed.
Solar Energy Intelligence Report, Oct. 22, 1979. Business Publishers, Inc., Silver Spring, Md., A. L. Frank, ed.
Solar Energy Intelligence Report, Nov. 26, 1979. Business Publishers, Inc., Silver Spring, Md., A. L. Frank, ed.
Stone, C. D., 1972. *Should Trees Have Standing?/Toward Legal Rights for Natural Objects*, William Kaufmann, Inc., Los Altos, Calif.
Thornton, T., 1978. *An Assessment of Anaerobic Digestion in U.S. Agriculture*, ESCS-06 PB-278-99918G1, Economics, Statistics, and Cooperatives Service, U.S. Department of Agriculture, Washington, D.C.
Tinberg, C., and A. D. McElroy, 1978. "The Economics of Growing Grasses and Grains for Energy," pp. 859–72 in *Proceedings of the Second Annual Symposium on Fuels from Biomass*, vol. II, *Rensselaer Polytechnic Institute, New York,* U.S. Department of Energy, Washington, D.C.
U.S. Department of Agriculture, 1978. *Gasohol from Grain—The Economic Issues*, ESCS-11, Accession No. PB28012, Washington, D.C.
U.S. Department of Energy, 1978. *Biomass Energy Success Stories—A Portfolio Illustrating Current Economic Uses of Renewable Biomass Energy*, Rep. HCP/T0285-01.
VanderWerf, P., 1978. *Regional Employment and Income Effects of a 50-MW Wood-Fired Power Plant*, Rep. DSD 90, prepared for Brookhaven National Laboratory under Contract 417613-S, Dartmouth College, Hanover, N.H.
Varani, F. T., J. Burford, and R. P. Arber, 1977. "The Design of a Large-Scale Manure/Methane Facility," in *Fuels and Energy from Renewable Resources*, D. A. Tillman et al., eds., Academic Press, New York.
Vergara, W., and D. Pimentel, 1978. "A Study of the Energy Potential of Fuels from Land Biomass in Five Countries," in *Symposium Papers—Energy from Biomass and Wastes, Washington, D.C., April 1978*, Institute of Gas Technology, Chicago, Ill.
Ward, B., 1973. "Only One Earth," in *Who Speaks for Earth?* W. W. Norton & Company, Inc., New York.
Wilcox, H. A., ed., 1976. *Ocean Food and Energy Farm Project*, vols. 1-7, ERDA/USN/1027-76, prepared for U.S. Energy Research and Development Administration by U.S. Naval Undersea Center, San Diego, Calif.
Yang, V., and S. C. Trinidade, 1978. "The Brazilian Gasohol Program" in *Symposium Papers—Energy from Biomass and Wastes, Washington, D.C., April 1978*, Institute of Gas Technology, Chicago, Ill.

Index

Applied General Equilibrium Modelling:

Applications, Limitations and Future Development

by

Professor D. (
Dr S. J. Leyb(
Mr G. V. Ree
University of

Professor J. W
University of

London: HMSO

First published 1993

ISBN 011 560049 3